Lecture Notes in Artificial Intelligence 6810

Edited by R. Goebel, J. Siekmann, and W. Wahlster

Subseries of Lecture Notes

FoLLI Publications on Logic, Language and Information

Łukasz Kaiser

Logic and Games
on Automatic Structures

Playing with Quantifiers and Decompositions

 Springer

Series Editors

Randy Goebel, University of Alberta, Edmonton, Canada
Jörg Siekmann, University of Saarland, Saarbrücken, Germany
Wolfgang Wahlster, DFKI and University of Saarland, Saarbrücken, Germany

Author

Łukasz Kaiser
LIAFA (CNRS)
Université Paris Diderot - Paris 7
175, Rue du Chevaleret
75013 Paris, France
E-mail: kaiser@liafa.jussieu.fr

ISSN 0302-9743   e-ISSN 1611-3349
ISBN 978-3-642-22806-3   e-ISBN 978-3-642-22807-0
DOI 10.1007/978-3-642-22807-0
Springer Heidelberg Dordrecht London New York

Library of Congress Control Number: 2011932847

CR Subject Classification (1998): F.4.1, I.2.3, I.1, D.2.4

LNCS Sublibrary: SL 7 – Artificial Intelligence

Typesetting: Camera-ready by author, data conversion by Scientific Publishing Services, Chennai, India

Printed on acid-free paper

Springer is part of Springer Science+Business Media (www.springer.com)

Foreword

Since 2002, FoLLI, the Association for Logic, Language, and Information (www.folli.org), has awarded an annual prize for an outstanding dissertation in the fields of logic, language, and information.

The prize is named after the well-known Dutch logician Evert Willem Beth, whose interdisciplinary interests are in many ways exemplary of the aims of FoLLI. It is sponsored by the E.W. Beth Foundation. Dissertations submitted for the prize are judged on technical depth and strength, originality, and impact made in at least two of the three fields of logic, language, and computation. Every year the competition is strong and the interdisciplinary character of the award stimulates lively debate in the Beth Prize Committee.

Recipients of the award are offered the opportunity to prepare a book version of their thesis for publication in the *FoLLI Publications on Logic, Language and Information.*

This volume is based on the PhD thesis of Łukasz Kaiser, who was a joint winner of the E.W. Beth dissertation award in 2009.

We wish to quote here the Committee's motivation for co-awarding the prize to him.

"Łukasz Kaiser's thesis on 'Logic and Games on Automatic Structures' is a very rich, technically highly involved, and innovative study in the area of algorithmic model theory, demonstrating the deep interplay between logic and computability in automatic structures.

In his thesis Dr. Kaiser solves several open problems, some of them in a surprising way and with very original ideas. In particular, he shows that first-order logic extended with regular game quantifiers is decidable in automatic structures and develops model-checking games for automatic structures. He also characterizes completely the unary generalized Lindström quantifiers that preserve regularity of relations in all omega-automatic structures, inter alia showing that all countable omega-automatic structures are in fact finite-word automatic.

Further, he proves the definability of the infinity and uncountability set quantifiers in MSO over countable linear orders and over labelled binary trees.

The thesis of Łukasz Kaiser displays very high technical and presentational quality, depth, originality, and rigor. It advances significantly the field of algorithmic model theory and raises interesting new questions, thus emerging as a fruitful and inspiring source for future research."

Valentin Goranko
(Chair of the Beth Prize Committee in 2009)
Michael Moortgat
(President of the Association for Logic, Language, and Information)

Preface

An important connection between logic and games is based on the correspondence between the evaluation of a logical formula and a game played by two opponents, one trying to show that the formula is true and the other trying to prove it false. This relationship has been implicitly known for a long time, even before mathematical logic and game theory were formalized. It was formally established in the 1950s by Paul Lorenzen [61, 62, 63] in the form of dialogue games and later developed in another form by Jaakko Hintikka [43]. Since then, it has inspired numerous research directions, leading both to new logics and interesting insights about the classical ones.

In computer science, there are two main approaches to algorithmically exploiting the correspondence between logic and games. On the one hand, games played on syntactic objects such as formulas, programs, or language expressions were studied. Such games, derived from the dialogue games of Lorenzen and their extensions, were used to build theorem provers for classical and intuitionistic first-order logic [30], to give semantics to programming languages and to verify programs [1], and in various other contexts in linguistics and artificial intelligence (see [65] for an overview). On the other hand, games can be played in a more semantic setting, where players choose elements of a mathematical structure. In this way, following the ideas of Hintikka, games are used to evaluate formulas of both first-order and second-order logic on finite structures and to verify temporal properties on Kripke structures.

The algorithmic utility of such semantic games is apparent in the verification of μ-calculus formulas on finite structures. While there is no known polynomial-time algorithm for this problem, parity and mean-payoff games were used to narrow its complexity class [47], to obtain algorithms that are among the most efficient ones in practice [48, 86], and recently to find the first sub-exponential algorithm for the verification of μ-calculus [49].

Extending the game-based algorithmic approach to first-order logic on *infinite* structures that arise in computer science is the main motivation for this thesis. In structures that are to be stored and manipulated by a computer, even infinite ones, elements and relations must be represented in a finite way.

For example, elements can be defined in an inductive way using algebraic datatypes, and relations can be given by programs that compute them. To avoid the problems inherent in theorem-proving with mathematical induction [20], we focus on the semantic setting where games are played using representations of elements of the structure. Since we are interested in algorithmic results, we additionally restrict our consideration to one prominent class of finitely presentable structures that has a decidable first-order theory, namely, to *automatic structures*.

Automatic structures, introduced first in [45] and later in [53] and [13], contain elements represented by words over a finite alphabet. Relations in these structures are represented by synchronous automata that perform step-by-step transitions on tuples of symbols from the alphabet. A prominent example of an automatic structure is Presburger arithmetic $(\mathbb{N}, +)$, for which the natural way of writing numbers as sequences of digits and the standard column addition method constitute an automatic presentation. In this thesis we allow words that represent elements of an automatic structure to be infinite; such structures are sometimes called ω-automatic.

The basic fact that first-order logic is decidable on automatic structures follows from the closure properties of automata, both the ones working on finite and those on infinite words [17]. To develop a correspondence between games and logic on automatic structures, we first look for suitable extensions of first-order logic that remain decidable on this class of structures. We study the notion of game quantification and extend the open and closed game quantifiers, known in model theory of infinitary logics, to a regular game quantifier defined on automatic presentations.

This quantifier corresponds to a construction of the words representing elements of a structure by means of a game played step-by-step with the letters from the alphabet. In this way the game quantifier intuitively captures games played by two players during the construction of elements of an automatic structure. We show that this quantifier effectively preserves regularity, which is closely related to the fact that alternating ω-automata can be determinized.

We study the expressive power of the regular game quantifier. We identify classes of structures on which logic extended with this quantifier collapses to pure first-order logic and distinguish these from those on which it has larger expressive power. We prove that quite basic structures, for example, the binary tree, are already complete for first-order logic extended with the game quantifier. To get a better understanding of the expressive power of this extended logic on weaker structures, we identify a class of inductive automorphisms and show that these preserve relations defined using the game quantifier.

Model-checking games for the extension of first-order logic with the game quantifier on automatic presentations can be defined in a more natural way than for pure first-order logic. To introduce them, we first recall the classical two-player parity games, which arise as the model-checking games for modal μ-calculus. We extend parity games to the multiplayer setting where two coalitions play against each other with a special kind of *hierarchical* imperfect information about actions of the players. This extension allows us to define the appropriate model-checking games for first-order logic with the regular game quantifier.

We look closely at the definition of hierarchical games to identify the influence of various factors on algorithmic properties of these games. On the one hand, it is essential to assume that the information is hidden in a hierarchical way and that players take moves in a prescribed alternating order. We show that allowing non-alternating moves of players makes the problem of determining the winners of these games undecidable. On the other hand, hierarchical games are robust under manipulations of the kind of winning condition in the game. The winning condition can be represented by a list of sets of positions which must appear infinitely often for one coalition to win (the Muller winning condition), by assigning priorities to positions and requiring that the minimal priority that appears infinitely often is even (the parity condition), or in other forms (e.g., the Streett and Rabin conditions). The complexity of establishing the winner in hierarchical games is not significantly affected by the kind of winning condition, as far as it is an ω-regular set of paths through the game graph.

One reason for the robustness of these games under changes of the winning condition is that adding a finite memory to strategies suffices to reduce games with complex winning conditions to games with easier ones. For example, it is a well-known result that games with Muller winning condition can be reduced to games with parity winning condition using finite memory. Gurevich and Harrington proved this in [39] using a special kind of memory structure called the latest appearance record, which follows the idea of order vectors introduced by McNaughton. Later, Zielonka introduced split trees [88], which form another memory structure that allows one to reduce Muller games to parity games and gives more insight into the amount of memory that is needed for the reduction when the Muller condition is fixed.

While these results are well-known for games over finitely many priorities, it has not been known how to extend these memory structures to games on infinite arenas with infinitely many priorities. We generalize the latest appearance record to a memory structure that can store a finite number of priorities that appeared in the play. Memory structures of this kind are sufficient for winning Muller games with a finite or co-finite number of sets in the Muller condition, and additionally for a few other classes of games with infinitely many priorities. Zielonka trees can be extended to certain classes of games with infinitely many priorities as well. We investigate these memory structures and show that the reductions known for the case of finitely many priorities can be generalized to games with infinitely many priorities, assuming certain constraints on the structure of the Zielonka tree for the winning condition.

Another common direction in which first-order logic can be extended is by adding generalized Lindström quantifiers [60]. This extension has been widely studied both in logic and in descriptive complexity theory, where it is used to describe complexity classes for machines with oracles [33]. We address the following question: which generalized unary quantifiers can be added to first-order logic without introducing non-regular relations on automatic structures? We answer this question with a complete characterization of such quantifiers. These are the cardinality and modulo counting quantifiers, i.e., "there exist infinitely many," "there exist uncountably many," and "there exist k mod m many." We show that these quantifiers indeed preserve regularity on all automatic structures, including the non-injective ω-automatic ones. As a corollary, we answer a question of Blumensath [11] and prove that all countable ω-automatic structures are in fact finite-word automatic.

A natural and often considered question is which other classes of structures still have decidable first-order logic with extensions. To investigate it, we study a large class of structures introduced by Colcombet and Löding [22], called generalized-automatic structures. These structures are given by interpretations transforming their first-order theory to monadic second-order theory of a tree or of a linear order with additional labels. Because of these arbitrary labellings, the methods from the previous chapters cannot be directly generalized. Instead, we use the composition method for monadic second-order theory over linear orders and trees.

Using the composition method we show that the second-order cardinality quantifiers, both the infinity and the uncountability quantifier on the number of sets X satisfying a formula $\varphi(X)$, can be effectively eliminated from monadic second-order logic. We devote one chapter to proving the elimination result over various linear orders, including all countable ones, and in the next chapter we prove the result for trees. This elimination procedure can be transferred to first-order logic only on injectively presented generalized-automatic structures and is thus a partial generalization of the results from previous chapters. It also illustrates how techniques from logic can be applied directly to automatic structures. We discuss the outlook on further extensions and applications of our work in the final chapter.

Acknowledgments. I wish to express my deep gratitude to Erich Grädel, not only for giving me the opportunity to meet inspiring people and to develop my interests, but also for his support and focus on the quality of scientific work. I am also grateful to Wolfgang Thomas for his encouragement and help.

I would like to thank Alexander Rabinovich and Sasha Rubin, I have gained much from their fascinating ways of thinking. I also thank Diana Fischer, Sophie Pinchinat, Dietmar Berwanger, Michael Ummels, and Tobias Ganzow for their illuminating comments and remarks which have contributed to this thesis, and for carefully correcting this text. I am especially grateful to my brother-in-arms Vince Bárány for his insightful thoughts, thorough discussions, and for his uncommon companionship during the years in Aachen.

For making my time in Aachen so enjoyable, I thank my friends Alex, Antje, Michaela, Christof, Jan, Frank, Kari, Michael, Nico, Philipp, Stefan, and also Bernd, Roman, and Wenyun. My deepest gratitude goes to my parents, Andrzej and Teresa, for their unending love, trust, and encouragement.

Contents

1

Logics, Structures and Presentations

In this chapter we review the standard notions of first-order and monadic second-order logic. We introduce linear orders and trees and state a few elementary properties of these structures. We recall the correspondence between monadic second-order logic and automata on infinite words together with basic facts from automata theory . We introduce automatic structures using presentations by automata and characterize them both by first-order and by monadic second-order to first-order interpretations. Finally, we discuss the composition method for monadic second-order logic over linear orders and trees.

As most of these notions are standard, we assume that the reader is already familiar with them and thus we do not discuss them in detail, but concentrate instead on fixing the notation. More thorough introductions to logic and automata can be found in many textbooks and surveys, e.g. in [27, 82, 84].

1.1 First-Order and Monadic Second-Order Logic

A structure $\mathfrak{A} = (A, R_1, \ldots, R_n, f_1, \ldots, f_m)$ is given by a set A, called the domain of $\mathfrak{A}$, a number of relations R_i and a number of functions f_j. If we denote the arity of R_i by r_i and the arity of f_j by s_j then $R_i \subseteq A^{r_i}$ and $f_j : A^{s_j} \to A$. We say that $\mathfrak{A}$ is a *relational structure* if it contains no functions, only relations. Every structure can be coded as a relational structure by replacing each function f_j by its graph G_{f_j}, which is a relation of arity $s_j + 1$ such that $G_{f_j}(\overline{x}, y) \iff f_j(\overline{x}) = y$. The signature of the structure $\mathfrak{A}$ is denoted by $\sigma(\mathfrak{A}) = \{R_i^{(r_i)}\} \cup \{f_j^{(s_j)}\}$, where R_i and f_j are now just symbols with appropriate arities. Note that we only consider structures with finite signatures in this thesis, even though some of the results do not rely on this assumption.

Both first-order and monadic second-order logic formulas over a signature τ are built from atomic formulas using Boolean connectives and quantifiers. Atomic formulas in first-order logic (FO) have either the form $t_1 = t_2$ or

$R_i(t_1, \ldots, t_{r_i})$, where t_k are terms, i.e. either first-order variables, denoted $x, y, x_1, y_1, \ldots$, or expressions of the form $f_j(t_1, \ldots, t_{s_j})$. In monadic second-order logic (MSO) there are additional atomic formulas of the form $t \in X$, where t is again a term and X is a second-order variable. We use the standard Boolean connectives $\wedge, \vee, \neg$ to denote conjunction, disjunction and negation and we write $\varphi \rightarrow \psi$ for $(\neg \varphi) \vee \psi$ and $\varphi \leftrightarrow \psi$ for $(\varphi \rightarrow \psi) \wedge (\psi \rightarrow \varphi)$. In first-order logic it is possible to quantify over first-order variables using either existential or universal quantifiers, i.e. to write $\exists x \varphi$ or $\forall x \varphi$. In second-order logic there is the additional possibility to quantify over second-order variables, i.e. to write $\exists X \varphi$ or $\forall X \varphi$. When writing formulas we use the convention that negation and quantifiers bind stronger than $\wedge$ and $\vee$, which in turn bind stronger than $\rightarrow$ and $\leftrightarrow$, so that $\forall x Rx \wedge \neg Ry \rightarrow Rz$ is to be understood as $((\forall x Rx) \wedge (\neg Ry)) \rightarrow Rz$. We say that a variable in a formula φ that appears in scope of a quantifier is *bound* and in the other case it is *free*. When writing $\varphi(x_1, \ldots, x_n)$ (or sometimes shorter $\varphi(\overline{x})$) we mean that all free variables of φ are contained in $\{x_1, \ldots, x_n\}$.

Whether a formula $\varphi(\overline{x}, \overline{X})$ holds in a structure $\mathfrak{A}$, given an assignment $\theta : \overline{x} \rightarrow A$ for first-order variables and an assignment $\Theta : \overline{X} \rightarrow \mathcal{P}(A)$ for second-order ones, denoted $\mathfrak{A}, \theta, \Theta \models \varphi$, is defined inductively. First, we extend the assignment θ to all terms by putting $\theta(f_j(t_1, \ldots, t_{s_j})) = f_j(\theta(t_1), \ldots, \theta(t_{s_j}))$. Note that f_j on the left side of the equation is only a symbol, while on the right side it is a function of $\mathfrak{A}$. We will often abuse the notation in this way and we sometimes write $f_i^{\mathfrak{A}}$ to point out that we have the function (or relation) in the structure in mind, rather than the symbol. Moreover, we extend this notation to formulas, so given a formula $\varphi(\overline{x})$ we write $\varphi^{\mathfrak{A}}$ for the relation defined by $\varphi^{\mathfrak{A}}(\overline{a}) \iff \mathfrak{A}, \overline{x} \leftarrow \overline{a} \models \varphi$. In the inductive definition of the semantics below we write $\theta[x \leftarrow a]$ (or analogous for Θ) for the assignment $\theta' : \overline{x} \cup \{x\} \rightarrow A$ that maps x to a and every other variable x' to $\theta(x')$.

- $\mathfrak{A}, \theta, \Theta \models t_1 = t_2$ whenever $\theta(t_1) = \theta(t_2)$,
- $\mathfrak{A}, \theta, \Theta \models R_i(t_1, \ldots, t_{r_i})$ whenever $R_i^{\mathfrak{A}}(\theta(t_1), \ldots, \theta(t_{r_i}))$ holds,
- $\mathfrak{A}, \theta, \Theta \models t \in X$ whenever $\theta(t) \in \Theta(X)$,
- $\mathfrak{A}, \theta, \Theta \models \neg \varphi$ whenever it is not the case that $\mathfrak{A}, \theta, \Theta \models \varphi$,
- $\mathfrak{A}, \theta, \Theta \models \varphi \wedge \psi$ whenever $\mathfrak{A}, \theta, \Theta \models \varphi$ and $\mathfrak{A}, \theta, \Theta \models \psi$,
- $\mathfrak{A}, \theta, \Theta \models \varphi \vee \psi$ whenever $\mathfrak{A}, \theta, \Theta \models \varphi$ or $\mathfrak{A}, \theta, \Theta \models \psi$,
- $\mathfrak{A}, \theta, \Theta \models \exists x \varphi$ whenever $\mathfrak{A}, \theta[x \leftarrow a], \Theta \models \varphi$ for some $a \in A$,
- $\mathfrak{A}, \theta, \Theta \models \forall x \varphi$ whenever $\mathfrak{A}, \theta[x \leftarrow a], \Theta \models \varphi$ for all $a \in A$,
- $\mathfrak{A}, \theta, \Theta \models \exists X \varphi$ whenever $\mathfrak{A}, \theta, \Theta[X \leftarrow B] \models \varphi$ for some $B \subseteq A$,
- $\mathfrak{A}, \theta, \Theta \models \forall X \varphi$ whenever $\mathfrak{A}, \theta, \Theta[X \leftarrow B] \models \varphi$ for all $B \subseteq A$.

Sometimes we evaluate MSO formulas in the *weak semantics* and in such case only finite sets are substituted for second-order variables. In this setting the last two items above must be replaced by the following:

- $\mathfrak{A}, \theta, \Theta \models \exists X \varphi$ in the weak semantics if $\mathfrak{A}, \theta, \Theta[X \leftarrow B] \models \varphi$ for some *finite* $B \subseteq A$,

  – $\mathfrak{A}, \theta, \Theta \models \forall X \varphi$ in the weak semantics if $\mathfrak{A}, \theta, \Theta[X \leftarrow B] \models \varphi$ for all *finite* $B \subseteq A$.

We often call formulas that we evaluate using the weak semantics *weak monadic second-order logic* (WMSO) formulas.

1.2   Linear Orders, Words and Trees

1.2.1   Linear Orders

Linear orders are a prominent example of structures that appear throughout this thesis. The standard ordering of natural numbers is denoted $(\omega, <)$ or $(\mathbb{N}, <)$, the orderings of integers and rational numbers are denoted $(\mathbb{Z}, <)$ and $(\mathbb{Q}, <)$, respectively. Recall that $(\mathbb{Q}, <)$ is dense, meaning that between any two elements $x < y$ there is another element z such that $x < z < y$. On the other hand, we say that a linear order $(L, <)$ is *scattered* if it does not embed any dense order, or equivalently if it does not embed $(\mathbb{Q}, <)$. Given a linear order $(I, <)$ and orders $(L_i, <_i)$ for every $i \in I$, the sum $\sum_I L_i$ is defined as the linear ordering of $\bigcup_{i \in I} L_i \times \{i\}$ such that

$$(l, i) < (l', i') \iff i < i' \text{ or } i = i' \text{ and } l <_i l'.$$

We write $(L_0, <_0) + (L_1, <_1)$ for the sum over $(\{0, 1\}, <)$. For example $(\mathbb{Z}, <)$ is isomorphic to $\omega^* + \omega$, where ω^* is the standard ordering of negative integers, $(\{-1, -2, -3, \ldots\}, <)$. Note that sometimes we use the same name for the structure and its universe when the meaning is clear from the context, as in the case of ω above. Linear orders with additional unary predicates are called *chains*. For a linear order L and $a, b \in L$ we use the standard notation for intervals, so for example $[a, b)$ is a left-closed and right-open interval. Moreover, we write $L|_{[a,b)}$ for the order $L \cap [a, b)$, and for $X \subseteq L$ we use analogous notation, i.e. $X|_{[a,b]}$ for $X \cap [a, b]$.

One can classify countable linear orders using the sum operation defined above in the following way. Every countable linear order can be written as a dense sum of scattered linear orders, i.e. as $\sum_{\mathbb{Q}} L_i$ where each $(L_i, <_i)$ is a scattered linear order. Moreover, Hausdorff classified countable scattered linear orders in classes VD_α defined inductively as follows. $\mathsf{VD}_0 = \{1\}$ consists of the linear order having one element (we leave out the empty linear order). For each ordinal $\alpha > 0$, VD_α consists of those linear orders that can be written as a sum $\sum_{\mathbb{Z}} L_i$ with each $L_i \in \mathsf{VD}_\beta$ for some $\beta < \alpha$. Let VD be the union of all the VD_α. Hausdorff has shown that VD contains every countable scattered linear order, and the *Hausdorff-rank* of a linear order $L \in \mathsf{VD}$ is defined as the smallest α such that $L \in \mathsf{VD}_\alpha$.

A linear order is *complete* if every one of its subsets has a least upper bound. Given a linearly ordered set $(L, <)$ its *Dedekind cuts* are subsets $C \subseteq L$ that are downward closed. The *completion* of $(L, <)$ is the set $DC(L)$ of Dedekind cuts of L ordered by inclusion, containing $\emptyset$ if there is no least element in L

and excluding $\emptyset$ in the other case. Note that the completion of L, which we denote by $\overline{L}$ and consider only up to isomorphism, is a complete linear order with both endpoints, i.e. a least and a greatest element.

Every linear order is naturally equipped with the *order topology* generated by open intervals. This allows us to speak of neighborhoods, open sets, limit (alternatively condensation or accumulation) points, and all other topological notions on every linear order.

1.2.2 Words

For a given set A we denote by A^* the set of all finite sequences of elements of A, by A^ω the set of all infinite sequences of elements of A (i.e. functions $\omega \rightarrow A$), and $A^{\leq\omega} = A^* \cup A^\omega$. Elements of A^* are often called finite words and elements of A^ω are infinite words over A. For any sequence $s = s_0 s_1 s_2 \ldots \in A^{\leq\omega}$ we denote by $|s|$ the length of s (either a natural number or ω) and by $s|_n = s_0 \ldots s_{n-1}$ the finite sequence composed of the first n elements of s, with $s|_0 = \varepsilon$, the empty sequence. We write $s[n]$ for the $(n+1)$st element of s (as we start counting from 0), so $s[n] = s_n$ for $n \in \mathbb{N}$. Similarly, $s[n, m]$ is the factor $s[n]s[n+1] \cdots s[m]$ and $s[n, m)$ is defined as $s[n, m-1]$, therefore in our notation $s|_n = s[0, n)$.

Given a finite sequence s and a sequence $r \in A^{\leq\omega}$ we denote by $s \cdot r$ (or just sr) the concatenation of s and r. For the n-times concatenation $s \cdots s$ we use the symbol s^n. Moreover, we write $s \sqsubseteq t$ if s is a prefix of t, i.e. if there exists a sequence r such that $t = sr$, and in such case we denote the difference by $t - s = r$. A subset B of $A^{\leq\omega}$ is said to be prefix-closed if for every $t \in B$ and $s \sqsubseteq t$ it holds that $s \in B$. For an infinite sequence $s \in A^\omega$ the set of elements that appear infinitely often in this sequence is denoted by $\mathrm{Inf}(s)$.

We sometimes extend all notations introduced above to vectors of sequences, so for example if $\bar{s}$ is a vector then $\overline{s[n]}$ or equivalently $\bar{s}[n]$ is the vector consisting of the $(n+1)$st element of each sequence in $\bar{s}$. Moreover, given a function $f : A \rightarrow B$ and $u \in A^{\leq\omega}$ we denote by $f(u)$ the sequence $f(u[0])\, f(u[1])\, f(u[2]) \ldots \in B^{\leq\omega}$.

1.2.3 Trees

A tree is a structure $\mathfrak{T} = (T, <, P_1, \ldots, P_n)$ where P_i's are unary predicates and $<$ is an irreflexive and transitive binary *ancestor* relation with a least element called the *root of* $\mathfrak{T}$ and such that for every $v \in T$ the set $\{u \in T \mid u < v\}$ of ancestors of v is finite and linearly ordered by $<$. We consider only finitely branching trees, i.e. we assume that in every tree $\mathfrak{T}$ the number of $v \in T$ with at most n ancestors is finite for every n.

Elements of a tree are referred to as *nodes*, maximal linearly ordered sets of nodes are called *branches*, ancestor-closed and linearly ordered sets of nodes are called *paths*, whereas *chains* are arbitrary linearly ordered subsets. An *antichain* is a set of pairwise incomparable nodes. Given a node v, the subtree of $\mathfrak{T}$ rooted in v is obtained by restricting the structure to the domain $T_v = \{u \in T \mid u \geq v\}$ and is denoted $\mathfrak{T}_v$. Similarly, we use this notation for every set $X \subseteq T$, i.e. $X_v = X \cap T_v$.

Given a finite set A we denote by $\mathfrak{T}(A)$ the complete tree over A, which is a structure with the universe A^* and $<$ interpreted as the standard prefix ordering. The tree $\mathfrak{T}(A)$ also includes successor labels, i.e. for each $a \in A$ there is a predicate P_a in the structure $\mathfrak{T}(A)$ such that $P_a(u)$ holds exactly when $u = va$, which allows to distinguish all immediate successors of any node $v \in T$. An important example of such a tree is the complete binary tree, $\mathfrak{T}(\{0,1\}) = (\{0,1\}^*, <, S_0, S_1)$ where $S_0(u)$ holds if $u = v0$ and $S_1(u)$ if $u = v1$. In the case of complete k-ary trees we may write $\mathfrak{T}(k)$ for $\mathfrak{T}(\{0, \ldots, k-1\})$, so the complete binary tree is designated $\mathfrak{T}(2)$.

Given an indexed family of trees and an order on the index which also satisfies the requirements of a tree, we define the sum of this family. Intuitively, the sum can be understood as replacing each node in the index tree by the corresponding tree in the family, and is formally defined as follows.

Definition 1.1 (Tree sum). *Let* $\mathfrak{I} = (I, <^{\mathfrak{I}})$ *be an unlabeled tree and for each* $i \in I$ *let* $\mathfrak{T}_i = (T_i, <^i, P_1^i, \ldots, P_n^i)$ *be a tree. The* tree sum *of* $(\mathfrak{T}_i)_{i \in \mathfrak{I}}$, *denoted* $\sum_{i \in \mathfrak{I}} \mathfrak{T}_i$, *is the tree*

$$\mathfrak{T} = \Big(\bigcup_{i \in I} \{i\} \times T_i \ , \ <^{\mathfrak{T}}, \bigcup_{i \in I} \{i\} \times P_1^i, \ldots, \bigcup_{i \in I} \{i\} \times P_n^i \Big),$$

where $(i, a) <^{\mathfrak{T}} (j, b)$ *for* $i, j \in I$, $a \in T_i$, $b \in T_j$ *iff:*

$i <^{\mathfrak{I}} j$ *and* a *is the root of* $\mathfrak{T}_i$, *or* $i = j$ *and* $a <^i b$.

Unless explicitly noted, we will not distinguish between $\mathfrak{T}_i$ *and the isomorphic subtree* $\{i\} \times \mathfrak{T}_i$ *of* $\mathfrak{T}$.

A particular special case of the sum we will be using is when the index structure $\mathfrak{I}$ consists of a single branch. Let $(I, <)$ be a linear order, which is finite or isomorphic to ω, and let $(\mathfrak{T}_i)_{i \in I}$ be an I-indexed sequence of trees. Then the sum $\mathfrak{T} = \sum_{i \in I} \mathfrak{T}_i$ is well defined, and $(I, <)$ forms a path (not necessarily maximal) in $\mathfrak{T}$.

In addition to the trees defined above, we also study trees over infinite words, called ω-trees. A complete ω-tree over a finite set A is defined in an analogous way to $\mathfrak{T}(A)$ as $\mathfrak{T}^\omega(A) = (A^{\leq \omega}, <, \{S_a\}_{a \in A})$, where $<$ is again the prefix relation and S_a are successor labels, i.e. $S_a(u)$ holds only for finite words $u = va$. Moreover, we sometimes extend the trees with the equal-length relation el, defined by $\mathsf{el}(u, w) \iff |u| = |w|$. We denote a tree $\mathfrak{T}$ extended with this binary relation $\mathfrak{T}_{\mathsf{el}}$, so for example $\mathfrak{T}^\omega_{\mathsf{el}}(\{0,1\})$ is the complete binary ω-tree extended with the equal-length relation, i.e. $(\{0,1\}^{\leq \omega}, <, S_0, S_1, \mathsf{el})$.

1.3 Automata on ω-Words

The order $(\omega, <)$ and the binary tree $\mathfrak{T}(2)$ play an important role in computer science and logic because the monadic second-order theory of both of these structures is decidable, as proved by Büchi [17] and Rabin [74] respectively. These proofs use the notion of an automaton, either a word automaton for $(\omega, <)$ or a tree automaton for $\mathfrak{T}(2)$, and establish a one-to-one correspondence between relations recognized by automata and the ones definable in monadic second-order logic.

An ω-word automaton $\mathcal{A}$ over a finite alphabet Σ is a tuple $(Q, \Delta, q_0, \mathcal{F})$ where Q is a finite set of states, Δ is a transition relation $\Delta \subseteq Q \times \Sigma \times Q$, $q_0 \in Q$ is an initial state and $\mathcal{F}$ is an acceptance condition. An automaton is *deterministic* if Δ is a function $Q \times \Sigma \to Q$. In the case of the standard finite-word automata, the acceptance condition consists only of a set of final states and a word is accepted if some run ends in a final state. For ω-word automata the acceptance condition $\mathcal{F}$ is a set of runs, i.e. infinite sequences of states, which are considered accepting for the automaton, so $\mathcal{F} \subseteq Q^\omega$. In practice $\mathcal{F}$ is described in a finite way and there are a few representations that are often used for this purpose.

- The *Büchi condition* is represented by a set $F \subseteq Q$ and
  $\mathcal{F} = \{s \in Q^\omega \mid \text{Inf}(s) \cap F \neq \emptyset\}$.
- The *parity condition* is defined using a mapping $\Omega : Q \to \{0, \dots, d\}$
  and $\mathcal{F} = \{s \in Q^\omega \mid \min(\text{Inf}(\Omega(s)))$ is even$\}$.
- The *Rabin condition* is given by a set of pairs $\{(E_1, F_1), \dots, (E_k, F_k)\}$
  and $\mathcal{F} = \{s \in Q^\omega \mid \text{Inf}(s) \cap F_i \neq \emptyset$ and $\text{Inf}(s) \cap E_i = \emptyset$ for some $i\}$.
- The *Streett condition*, dual to the Rabin condition, is again represented by
  a set of pairs $\{(E_1, F_1), \dots, (E_k, F_k)\}$, but in this case
  $\mathcal{F} = \{s \in Q^\omega \mid \text{Inf}(s) \cap F_i \neq \emptyset$ or $\text{Inf}(s) \cap E_i = \emptyset$ for every $i\}$.
- The *Muller condition* is defined by listing $F \subseteq \mathcal{P}(Q)$ and
  $\mathcal{F} = \{s \in Q^\omega \mid \text{Inf}(s) \in F\}$.

A *run* of an automaton $\mathcal{A}$ on a word $w \in \Sigma^\omega$ is defined as any sequence of states $q_0 q_1 \dots \in Q^\omega$ in which $\Delta(q_i, w[i], q_{i+1})$ holds for all i. The word w is *accepted* by $\mathcal{A}$ if there is a run ρ of $\mathcal{A}$ on w that is accepting, i.e. $\rho \in \mathcal{F}$, and we denote by $L(\mathcal{A})$ the set of all words accepted by an automaton $\mathcal{A}$.

It is well-known that non-deterministic Büchi, parity, Rabin, Streett and Muller automata all recognize the same class of languages, the ω-regular languages. The deterministic variants have the same expressive power for all the representations of acceptance conditions introduced above except for the case of Büchi condition, as deterministic Büchi automata are strictly weaker than non-deterministic ones. Moreover, the class of ω-regular languages is closed under union, intersection and complementation.

1.3.1 Alternating Automata

In addition to the standard notion of automata, we use *alternating* automata as a tool in our proofs. The intuition behind alternating automata is that, unlike in the deterministic case where only one run on a given word is possible, there are more possibilities of transitions from each state for a given letter. But unlike non-deterministic automata, an alternating automaton does not only accept a word if there *exists* an accepting run among all possible ones, or if *all* possible runs are accepting (as in universal automata), but it allows to alternate such conditions with respect to states of the automaton and has both existential and universal branching choices.

To define alternating automata we have to consider, for a given set of states Q, the set $\mathcal{B}^+(Q)$ of all *positive Boolean formulas* over Q. By definition $\mathcal{B}^+(Q)$ is the set of all Boolean formulas built using elements of Q, the Boolean connectives $\wedge$ and $\vee$ and the constants $\top$ (true) and $\bot$ (false). Note that negation is not allowed. We say that a subset $X \subseteq Q$ *satisfies* a formula $\varphi \in \mathcal{B}^+(Q)$ if φ is satisfied by the assignment that assigns true to all elements of X and false to $Q \setminus X$.

An *alternating automaton* $\mathcal{A}$ over an alphabet Σ is a tuple $(Q, \delta, q_0, \mathcal{F})$, where Q is the set of states, q_0 is the initial state, $\mathcal{F}$ is the acceptance condition, but this time δ does not point to a single next state but specifies a positive Boolean formula as transition condition, $\delta \; : \; Q \times \Sigma \to \mathcal{B}^+(Q)$. Intuitively, a *correct run* of $\mathcal{A}$ on a word w is a tree labeled with Q where the successors of each node form a satisfying set for the Boolean condition related to the state in this node and to the corresponding letter in w.

To capture this intuition formally and simplify notation, we define runs as sets of infinite sequences of states, so a run ρ is a subset of Q^ω. When one thinks of runs as trees, our definition corresponds to defining runs directly as the set of branches of the run-tree. For a run ρ represented in this way we write $s_\rho(u)$ for the set of all states appearing in ρ that prolong $u \in Q^*$, i.e. the successors of u when thinking of a run as a tree,

$$s_\rho(u) = \{q \in Q \; : \; \exists v \; u \cdot q \cdot v \in \rho\}.$$

We define that ρ is a correct run of $\mathcal{A}$ on the word w if for each infinite branch $u \in \rho$ and each prefix $u|_i$ the successors after that prefix satisfy the corresponding Boolean constraint, i.e. if $s_\rho(u|_i)$ satisfies $\delta(u[i], w[i])$ for all i and $u \in \rho$. We say that $\mathcal{A}$ accepts a word w if there is a correct, non-empty run ρ on w starting from q_0 such that each branch $u \in \rho$ is accepted, i.e. $u \in \mathcal{F}$, and again we write $L(\mathcal{A})$ for the language recognized by $\mathcal{A}$.

Alternating automata may seem more powerful than deterministic ones and it is often much easier to express problems in terms of alternating automata than in terms of deterministic ones, but they are in fact equal in expressiveness to the standard automata.

Theorem 1.2. *Every language recognized by an alternating Büchi, parity, Rabin, Streett or Muller automaton is ω-regular.*

The above theorem can be proved by expressing acceptance of alternating automata in monadic second-order logic on infinite words and then going back from the logic to automata [17]. Alternatively, one can give an explicit construction which shows that (for all acceptance conditions except the Büchi condition) the size of the deterministic automaton constructed for a language recognized by an alternating one is at most doubly exponential in the size of the original automaton, as first shown in [66].

1.3.2 ω-Semigroups

There is a fundamental correspondence between recognizability of sets by finite-word automata and by finite semigroups. It has been extended to recognizability of ω-regular sets, first using Wilke algebras [87] and later based on the notion of ω-semigroups. The theory of ω-semigroups was first well presented in [71] and is thoroughly discussed in [72], we only mention what is most necessary.

An ω-semigroup $S = (S_f, S_\omega, \cdot, *, \pi)$ is a two-sorted algebra, where $(S_f, \cdot)$ is a semigroup, $* : S_f \times S_\omega \mapsto S_\omega$ is the *mixed product* satisfying for every $s, t \in S_f$ and every $\alpha \in S_\omega$ the equality

$$s * (t * \alpha) = (s \cdot t) * \alpha$$

and where $\pi : S_f^\omega \mapsto S_\omega$ is the *infinite product* satisfying

$$s_0 * \pi(s_1, s_2, \ldots) = \pi(s_0, s_1, s_2, \ldots)$$

as well as the associativity rule

$$\pi(s_0, s_1, s_2, \ldots) = \pi(s_0 s_1 \cdots s_{k_1}, s_{k_1+1} s_{k_1+2} \cdots s_{k_2}, \ldots)$$

for every sequence $(s_i)_{i \geq 0}$ of elements of S_f and every strictly increasing sequence $(k_i)_{i \geq 0}$ of indices. For $s \in S_f$ we denote $s^\omega = \pi(s, s, \ldots)$.

Morphisms of ω-semigroups are defined to preserve all three products as expected. There is a natural way to extend finite semigroups and their morphisms to ω-semigroups. As in semigroup theory, idempotents play a central role in this extension. An *idempotent* is a semigroup element $e \in S$ satisfying $ee = e$. For every element s in a finite semigroup the sub-semigroup generated by s contains a unique idempotent s^k. The least $k > 0$ such that s^k is idempotent for every $s \in S_f$ is called the *exponent* of the semigroup S_f. Another useful notion is absorption of semigroup elements: we say that s *absorbs* t *(on the right)* if $st = s$.

There is a natural extension of the free semigroup Σ^+ to the free ω-semigroup $(\Sigma^+, \Sigma^\omega)$ with $*$ and π determined by concatenation. An ω-semigroup $S = (S_f, S_\omega)$ *recognizes* a language $L \subseteq \Sigma^\omega$ via a morphism $\phi : (\Sigma^+, \Sigma^\omega) \to (S_f, S_\omega)$ if $\phi^{-1}(\phi(L)) = L$. This notion of recognizability coincides, as for finite words, with recognizability by non-deterministic Büchi automata and translations from Büchi automata to ω-semigroups and back can be done effectively.

Theorem 1.3 ([71]). *A language $L \subseteq \Sigma^\omega$ is ω-regular if and only if it is recognized by a finite ω-semigroup.*

This correspondence allows one to engage in an algebraic study of varieties of ω-regular languages, and also has the advantage of hiding complications of cutting apart and stitching together runs of Büchi automata. This is precisely the reason for which we use this algebraic framework. Most remarkably, one does not need to understand the exact relationship between automata and ω-semigroups and the technical details of the constructions behind Theorem 1.3 to use ω-semigroups to simplify calculations on ω-regular sets.

1.4 Automatic Structures

Before we define automatic presentations and automatic structures, let us introduce relations on finite and ω-words recognized by ω-word automata operating in a synchronized letter-to-letter fashion. Formally, R is an ω-*regular relation* of arity r over the domain Σ^ω if there exists an ω-automaton $\mathcal{A}$ over the alphabet Σ^r accepting the *convolution* $\otimes \overline{w}$ of ω-words $w_1, \ldots, w_r$ exactly when $R(w_1, \ldots, w_r)$ holds. The convolution is defined as

$$\otimes \overline{w}[i] = (w_1[i], \ldots, w_r[i]) \text{ for all } i.$$

For pairs of words w, w' we sometimes write $w \otimes w'$ or $\binom{w}{w'}$ for $\otimes(w, w')$.

Example 1.4. Words $u, v \in \Sigma^\omega$ have *equal ends*, written $u \sim_e v$, if $u[n] = v[n]$ for all but a finitely many natural numbers n. This is an important example of an ω-regular equivalence relation. For $S, T \subseteq \mathbb{N}$ we extend the notation and write $S \sim_e T$ if for all but finitely many $n \in \mathbb{N}$, $n \in S \iff n \in T$. The non-deterministic Büchi automaton depicted in Figure 1.1 accepts the equal-ends relation over $\{0, 1\}$.

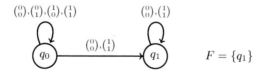

Fig. 1.1. An automaton for the equal ends relation

To define ω-regular relations over finite words one needs to add a padding end-of-word symbol $\square \notin \Sigma$ to formally define the convolution of words of different length. For simplicity, we will sometimes identify a finite word $w \in \Sigma^*$ with its infinite padding $w^\square = w\square^\omega \in \Sigma_\square^\omega$ where $\Sigma_\square = \Sigma \cup \{\square\}$. A language $L \subseteq \Sigma^*$ is regular, i.e. recognized by a standard finite-word automaton, exactly if the language $L^\square = \{w^\square \mid w \in L\}$ is ω-regular over $\Sigma_\square$. Thus, we say that an r-ary relation $R \subseteq (\Sigma^*)^r$ is *regular* whenever its extension with padding is ω-regular over $\Sigma_\square$. This is equivalent to defining a convolution for finite words by padding each word with $\square$ to be as long as the longest one.

Definition 1.5 (Automatic Presentation)
For any relational structure $\mathfrak{A} = (A, R_1, \ldots, R_k)$, a tuple of ω-automata $\mathfrak{d} = (\mathcal{A}, \mathcal{A}_\approx, \mathcal{A}_1, \ldots, \mathcal{A}_k)$ together with a surjective naming function $f : L(\mathcal{A}) \to A$ constitutes an $(\omega\text{-})$automatic presentation of $\mathfrak{A}$ over Σ if the following criteria are met:

(i) the equivalence relation denoted $\approx$ and defined as

$$\{(u, w) \in L(\mathcal{A})^2 \mid f(u) = f(w)\}$$

is recognized by $\mathcal{A}_\approx$,

(ii) every automaton $\mathcal{A}_i$ recognizes a relation $\mathcal{R}_i \subseteq (\Sigma^\omega)^{r_i}$ with the same arity r_i as the relation R_i,

(iii) f is an isomorphism between $\mathfrak{A}_\mathfrak{d} = (L(\mathcal{A}), \mathcal{R}_1, \ldots, \mathcal{R}_k)/_\approx$ and $\mathfrak{A}$.

The presentation is said to be injective *whenever f is, in which case $\mathcal{A}_\approx$ can be omitted. Observe that the relation $\approx$ needs to be a congruence of the structure $(L(\mathcal{A}), \mathcal{R}_1, \ldots, \mathcal{R}_k)$ for item (iii) to hold.*

In case $L(\mathcal{A})$ only consists of words of the form $w^\square$ where $w \in \Sigma^*$, we say that the presentation is (finite-word) automatic. We call a structure $(\omega\text{-})$*automatic* if it has an $(\omega\text{-})$automatic presentation.

There may exist different automatic presentations of a single structure, and different relations might be regular in each of these presentations. For example, for every number $p > 1$ there is a presentation of Presburger arithmetic $(\mathbb{N}, +)$ where numbers are coded in base p. The relation $|_2$ defined as $x|_2 y \iff x|y$ and $x = 2^n$ is a regular relation when numbers are coded in binary, but it is not regular in the presentation that uses ternary coding. Relations that are regular in each automatic presentation of a structure are called *intrinsically regular* and were first studied in [54, 55]. Every FO-definable relation is intrinsically regular, but on some structures there are intrinsically regular relations that are not definable in FO. Remarkably, this is not the case for Presburger arithmetic, where all relations that are intrinsically regular are definable in FO. An accessible survey of results on presentations of Presburger arithmetic is given in [16], and the general problem of different automatic presentations of a structure is studied in [5, 6].

The basic advantage of having an $(\omega\text{-})$automatic presentation of a structure lies in the fact that first-order formulas can be effectively evaluated using classical automata constructions. This is expressed by the following fundamental theorem.

Theorem 1.6 (Cf. [45, 53, 14]). *There is an effective procedure that given an $(\omega\text{-})$automatic presentation $\mathfrak{d}, f$ of a structure $\mathfrak{A}$, and given a FO-formula $\varphi(\overline{x})$ constructs an $(\omega\text{-})$automaton recognizing $f^{-1}(\varphi^\mathfrak{A})$. The FO-theory of every $(\omega\text{-})$automatic structure is decidable.*

1.5   Interpretations and Complete Structures

In this section, instead of explicitly representing a structure by a finite object, as done above using automata, we consider operations for transforming structures. More precisely, we fix an underlying family of structures and a class of operations that transform structures, and investigate the class of structures obtained by applying the transformations to structures in the underlying family. When the operations preserve decidability of first-order or monadic second-order logic, and structures in the underlying family have decidable first-order or monadic second-order theory, then we obtain a class of structures on which the corresponding logic is again decidable.

An important and well studied way of transforming structures is the model-theoretic interpretation, where one structure is interpreted in another one using formulas of either first-order or monadic second-order logic. Interpretations preserve decidability of the corresponding logics, or transform structures with decidable monadic second-order theory to structures with decidable first-order theory. We will show a few ways in which automatic structures can be characterized and extended by means of interpretations in trees and linear orders.

Definition 1.7. *Let $\mathfrak{A} = (A, R_1, \ldots, R_k)$ and $\mathfrak{B}$ be relational structures and let r_i be the arity of R_i. A tuple of first-order formulas over $\sigma(\mathfrak{B})$,*

$$\mathcal{I} = (\delta(\overline{x}), \varepsilon(\overline{x_1}, \overline{x_2}), \varphi_1(\overline{x_1}, \ldots, \overline{x_{r_1}}), \ldots, \varphi_k(\overline{x_1}, \ldots, \overline{x_{r_k}})),$$

where each vector $\overline{x}, \overline{x_i}$ is of the same length n is an n-dimensional FO *interpretation of $\mathfrak{A}$ in $\mathfrak{B}$ if $\mathcal{I}(\mathfrak{B}) = (\delta^{\mathfrak{B}}, \varphi_1^{\mathfrak{B}}, \ldots, \varphi_k^{\mathfrak{B}})/\varepsilon^{\mathfrak{B}}$ is isomorphic to $\mathfrak{A}$.*

In an analogous way, a tuple of MSO *or* WMSO *formulas over $\sigma(\mathfrak{B})$,*

$$\mathcal{J} = (\delta(X), \varepsilon(X_1, X_2), \varphi_1(X_1, \ldots, X_{r_1}), \ldots, \varphi_k(X_1, \ldots, X_{r_k})),$$

where X and X_i are single second-order variables, is an MSO-to-FO *or a* WMSO-to-FO *interpretation if $\mathcal{J}(\mathfrak{B}) = (\delta^{\mathfrak{B}}, \varphi_1^{\mathfrak{B}}, \ldots, \varphi_k^{\mathfrak{B}})/\varepsilon^{\mathfrak{B}}$ is isomorphic to $\mathfrak{A}$, with the formulas evaluated using the standard or the weak semantics respectively. If there exists an* FO, MSO-to-FO *or* WMSO-to-FO *interpretation of $\mathfrak{A}$ in $\mathfrak{B}$ we denote this by $\mathfrak{A} \leq_{\mathsf{FO}} \mathfrak{B}$, $\mathfrak{A} \leq_{\mathsf{MSO} \to \mathsf{FO}} \mathfrak{B}$ or $\mathfrak{A} \leq_{\mathsf{WMSO} \to \mathsf{FO}} \mathfrak{B}$, respectively.*

Let ψ be a first-order formula over $\sigma(\mathfrak{A})$ and $\mathcal{I}$ an interpretation of $\mathfrak{A}$ in $\mathfrak{B}$. We construct the formula $\psi^{\mathcal{I}}$ by replacing every relation symbol R_i in ψ by the corresponding formula φ_i of $\mathcal{I}$, replacing every equality $t_1 = t_2$ in ψ by $\varepsilon(t_1, t_2)$ and relativizing the quantifiers in the following way. In the case of FO interpretations we replace $\exists x \varphi$ by $\exists \overline{x}(\delta(\overline{x}) \wedge \varphi)$ and $\forall x \varphi$ by $\forall \overline{x}(\delta(\overline{x}) \to \varphi)$, and in the second-order case we use second-order variables and thus $\exists X(\delta(X) \wedge \varphi)$ and $\forall X(\delta(X) \to \varphi)$, respectively. The standard interpretation lemma states that $\mathfrak{A} \models \psi$ exactly if $\mathfrak{B} \models \psi^{\mathcal{I}}$. This allows us to deduce decidability of the

FO-theory of $\mathfrak{A}$ from decidability of the FO, MSO or WMSO theory of $\mathfrak{B}$, given an FO, MSO-to-FO or WMSO-to-FO interpretation of $\mathfrak{A}$ in $\mathfrak{B}$, respectively.

A class $\mathcal{K}$ of structures is *closed under* FO-*interpretations* if for all $\mathfrak{B} \in \mathcal{K}$ whenever $\mathfrak{A} \leq_{\mathsf{FO}} \mathfrak{B}$ then $\mathfrak{A} \in \mathcal{K}$ as well. For such a class $\mathcal{K}$ we say that a structure $\mathfrak{B}$ is *complete for* $\mathcal{K}$ if all $\mathfrak{A} \in \mathcal{K}$ are FO-interpretable in $\mathfrak{B}$. It follows from Theorem 1.6 that the class of (ω-)automatic structures is closed under FO-interpretations, because every relation defined in FO using only (ω-)regular relations is again (ω-)regular.

One way to characterize automatic structures by means of interpretations is to find complete automatic structures, and such structures were presented in [11, 14]. It was shown there that for any finite alphabet Σ with at least two letters, the complete tree over Σ extended with the equal-length relation, $\mathfrak{T}_{\mathsf{el}}(\Sigma)$, is complete for the class of automatic structures. Analogously, the complete ω-tree with the equal-length relation, $\mathfrak{T}_{\mathsf{el}}^{\omega}(\Sigma)$, is complete for the class of ω-automatic structures.

It is natural to ask whether Presburger arithmetic is a complete automatic structure. The answer is negative, but there are extensions of Presburger arithmetic that are complete. Let us define the following structures: $\mathfrak{N}_p = (\mathbb{N}, +, |_p)$ where $x|_p y \iff x|y$ and $x = p^n$ for some $n \in \mathbb{N}$, and $\mathfrak{R}_p = (\mathbb{R}, +, \leq, |_p, 1)$ where $x|_p y \iff y = kx$ and $x = p^l$ for some $k, l \in \mathbb{Z}$. It was shown in [11, 14] that for all integers $p \geq 2$ the extensions $\mathfrak{N}_p$ and $\mathfrak{R}_p$ of Presburger arithmetic and the real arithmetic are indeed complete, for the class of automatic and ω-automatic structures respectively.

Another way to characterize automatic structures, where WMSO-to-FO and MSO-to-FO interpretations are used, was first mentioned in [78] and more systematically introduced in [22]. This characterization extends the intimate connection between ω-automata over words and MSO over $(\omega, <)$, as well as between finite-word automata and WMSO over $(\omega, <)$, to (ω-)automatic structures. A structure $\mathfrak{A}$ is finite-word automatic if there is a WMSO-to-FO interpretation of $\mathfrak{A}$ in $(\omega, <)$, and a structure $\mathfrak{B}$ is ω-automatic if there is an MSO-to-FO interpretation of $\mathfrak{B}$ in $(\omega, <)$. Let us summarize the characterizations of automatic structures by means of interpretations in the following theorems.

Theorem 1.8 (Cf. [11, 14, 78, 22]). *For any relational structure $\mathfrak{A}$ the following statements are equivalent:*

- $\mathfrak{A}$ *is finite-word automatic,*
- $\mathfrak{A} \leq_{\mathsf{FO}} \mathfrak{T}_{\mathsf{el}}(2)$,
- $\mathfrak{A} \leq_{\mathsf{FO}} \mathfrak{N}_2$,
- $\mathfrak{A} \leq_{\mathsf{WMSO} \to \mathsf{FO}} (\omega, <)$.

Theorem 1.9 (Cf. [11, 14, 78, 22]). *For any relational structure $\mathfrak{A}$ the following statements are equivalent:*

- $\mathfrak{A}$ *is ω-automatic,*
- $\mathfrak{A} \leq_{\mathsf{FO}} \mathfrak{T}^{\omega}_{\mathsf{el}}(2)$,
- $\mathfrak{A} \leq_{\mathsf{FO}} \mathfrak{R}_2$,
- $\mathfrak{A} \leq_{\mathsf{MSO} \to \mathsf{FO}} (\omega, <)$.

The characterization of automatic structures by MSO-to-FO interpretations was used in [22] to define *generalized-automatic structures*. We say that $\mathfrak{A}$ is an $(\omega$-)generalized-automatic structure if there is a WMSO-to-FO (or MSO-to-FO) interpretation of $\mathfrak{A}$ in some tree $\mathfrak{T}$. In particular, we say that the structure is $(\omega$-)tree-automatic if this is the case for $\mathfrak{T}(2)$, the complete binary tree. By the result of Rabin and the interpretation lemma, $(\omega$-)tree-automatic structures have a decidable first-order theory. Moreover, in chapter 7 we show that certain extensions of first-order logic collapse to FO on all ω-generalized-automatic structures.

1.6   Composition in Monadic Second-Order Logic

To study logic on linear orders and trees with arbitrary additional predicates it is convenient to depart from automata and use related methods from mathematical logic, especially the composition method. The history of the composition method starts with the introduction of Ehrenfeucht games [29], which are an intuitive formulation of Fraïssé's characterization of elementary equivalence, i.e. indistinguishability of relational structures by first-order formulas. These games were first defined for first-order logic and extended to weak monadic second-order logic [29]. Later, other logical systems were covered, such as modal, temporal and infinitary logics that we discuss in chapter 2. Here we focus on the extension of this method, now usually called the Ehrenfeucht-Fraïssé method, to full monadic second-order logic over linear orders and trees.

While Ehrenfeucht proved decidability of the first-order theory of countable ordinals using logical methods [28, 29], decidability of the full monadic second-order theory of these orderings was first shown by Büchi using automata [17, 18, 19]. Only later Shelah gave, in his celebrated and difficult paper [80], alternative proofs of Büchi's results (and many more) using an extension of the Ehrenfeucht-Fraïssé method to full monadic second-order logic, which he called the composition of monadic theories. This method was subsequently used by Gurevich and Shelah to obtain even more results, for example in [37, 41] and with Magidor in [40]. Theoretical computer scientists long preferred the automata theoretic approach, even after the composition method was well presented in Gurevich's survey [38]. It was only after the more accessible survey by Wolfgang Thomas [83] that the merits of the composition method started to be appreciated in theoretical computer science, which resulted in numerous papers. One example is the characterization of all extensions of $(\omega, <)$ by unary predicates that have a decidable monadic second-order theory [75].

The quantifier rank of a formula φ, denoted $qr(\varphi)$, is the maximum depth of nesting of quantifiers in φ. For fixed n and l (and a fixed signature) we denote by $Form_{n,m}$ the set of formulas of quantifier depth $\leq n$ and with free variables among $X_1, \ldots, X_m$.

For a structure $\mathfrak{A}$ and a tuple $\overline{U}$ of m subsets of $\mathfrak{A}$, the *monadic n-theory of $\overline{U}$*, $Th^n(\mathfrak{A}, \overline{U})$, is the set of all MSO formulas $\varphi(\overline{X}) \in Form_{n,m}$, having no more than n nested quantifiers in any subformula and no free variables other than $X_1, \ldots, X_m$, for which $\mathfrak{A} \models \varphi(\overline{U})$, i.e.

$$Th^n(\mathfrak{A}, \overline{U}) = \{ \varphi(\overline{X}) \in Form_{n,m} \mid \mathfrak{A} \models \varphi(\overline{U}) \}.$$

For any $n, m > 0$, the set $Form_{n,m}$ is infinite, but it only contains finitely many semantically distinct formulas, i.e. there are only finitely many n-theories in m variables. Moreover, every n-theory $Th^n(\mathfrak{A}, \overline{U})$ is definable by a single MSO formula $\tau(\overline{X})$ having m free variables and quantifier depth at most n. Hintikka formulas are canonical formulas defining n-theories.

Lemma 1.10 (Hintikka Lemma [42]). *For every $n, m \in \mathbb{N}$ (and a fixed signature), we can compute a finite set $H_{n,m} \subseteq Form_{n,m}$ such that:*

- *For every structure $\mathfrak{A}$ and $\overline{U} \subseteq \mathfrak{A}$ there is a unique $\tau \in H_{n,m}$ such that $\mathfrak{A} \models \tau(\overline{U})$.*
- *If $\tau_1, \tau_2 \in H_{n,m}$ and $\tau_1 \neq \tau_2$ then $\tau_1 \wedge \tau_2$ is unsatisfiable.*
- *If $\tau \in H_{n,m}$ and $\varphi \in Form_{n,m}$, then either $\tau \models \varphi$ or $\tau \models \neg\varphi$. Furthermore, there is an algorithm that, given such τ and φ, decides which of these two possibilities holds.*

Elements of $H_{n,m}$ are called (n, m)-Hintikka formulas.

We say that a structure $\mathfrak{A}$ with labels (unary predicates) $\overline{U}$ has *type* $\tau \in H_{n,m}$, denoted $Tp^n(\mathfrak{A}, \overline{U}) = \tau$, if $\mathfrak{A} \models \tau(\overline{U})$, i.e. if τ and $Th^n(\mathfrak{A}, \overline{U})$ are equivalent. We sometimes speak of the n-type of a tuple of subsets $\overline{V} = V_1, \ldots, V_m$ of a given structure $\mathfrak{A}$ which already contains labels $\overline{U} = U_1, \ldots, U_l$. This is synonymous with the n-type $\tau \in H_{n,l+m}$ of the structure $(\mathfrak{A}, \overline{V})$ obtained by expansion of $\mathfrak{A}$ with the predicates interpreted as $\overline{V}$.

The essence of the composition method is that certain operations on structures, such as disjoint union and ordered sums of linear orders, can be projected to n-theories, i.e. there are corresponding operations mapping n-theories of constituent structures to the n-theory of the resulting structure. In other words, n-theories can be composed.

Here we state a very simple form of the composition method on linear orders and on trees, which can be proven directly using Ehrenfeucht-Fraïssé games. As mentioned before, there are more powerful theorems also known as the composition method, e.g. the effective ones presented later in chapters 6 and 7 and other, c.f. [80, 37, 41, 40].

Theorem 1.11 (Composition on linear orders)
Let $(I, <)$ be a linear order, and $\{\mathfrak{L}_i \mid i \in I\}$ and $\{\mathfrak{L}'_i \mid i \in I\}$ two I-indexed sequences of chains such that $\mathrm{Tp}^n(\mathfrak{L}_i) = \mathrm{Tp}^n(\mathfrak{L}'_i)$ for all $i \in I$. Then $\mathrm{Th}^n\left(\sum_{i \in I} \mathfrak{L}_i\right) = \mathrm{Th}^n\left(\sum_{i \in I} \mathfrak{L}'_i\right)$.

Theorem 1.12 (Composition on tree sums)
Let $\mathfrak{J} = (I, <^I)$ be a fixed unlabeled tree. For every family $\{\mathfrak{T}_i \mid i \in I\}$ of trees, the theory $\mathrm{Th}^n(\sum_{i \in \mathfrak{J}} \mathfrak{T}_i)$ is uniquely determined by the theories $\mathrm{Th}^n(\mathfrak{T}_i)$.

2

Game Quantifiers on Automatic Presentations

This chapter is devoted to the study of game quantification on automatic presentations. We start with a historic survey on game quantification in the context of infinitary logics, based on [56]. Then, we define a new game quantifier on automatic presentations, the regular game quantifier. We show that the basic advantage of first-order logic on automatic structures, namely decidability and regularity of definable relations, still holds for the logic extended with the game quantifier.

Further, we investigate the expressive power of game quantification on automatic presentations. We show that the prefix and equal-length relations on a presentation are definable using only equality and the regular game quantifier. It follows that much simpler structures are complete for the extended first-order logic than for the standard one. In contrast to the binary ω-tree with equal-length $\mathfrak{T}_{\mathsf{el}}^{\omega}(2)$ needed for pure first-order logic, already the binary ω-tree $\mathfrak{T}^{\omega}(2)$, even without the prefix relation, is complete for first-order logic extended with the game quantifier.

To understand which relations are not definable using the game quantifier, we study the automorphisms of structures that preserve formulas of the extended logic. For this reason we introduce the notion of inductive automorphisms and show that all relations definable in first-order logic extended with the regular game quantifier are preserved under such automorphisms. In addition to the explicit definition, we characterize inductive automorphisms as exactly those automorphisms that preserve the prefix relation on an automatic presentation.

2.1 Open and Closed Game Quantifiers

It is natural to ask how first-order logic can be extended without allowing second-order quantification. Two possible extensions are well studied: one where new unary quantifiers are allowed in addition to $\exists$ and $\forall$, which is discussed in more detail in section 5.1, and another where infinitely long formulas can be written, either by allowing infinite conjunctions and disjunctions or by allowing infinite strings of quantifiers.

The extension of first-order logic where conjunctions and disjunctions of size less than κ and *homogeneous* strings of quantifiers of length less than λ are allowed is denoted $\mathsf{L}_{\kappa\lambda}$ for any cardinals κ and λ. For example $\mathsf{FO} = \mathsf{L}_{\omega\omega}$, and the extension where only countable conjunctions and disjunctions are allowed is $\mathsf{L}_{\omega_1\omega}$. Allowing countable Boolean operations means that the syntax of $\mathsf{L}_{\omega_1\omega}$ allows to build formulas of the form $\bigwedge_{i\in\mathbb{N}} \varphi_i$ and $\bigvee_{i\in\mathbb{N}} \varphi_i$ in addition to what is allowed by the standard FO syntax. The definition of semantics for $\mathsf{L}_{\omega_1\omega}$ contains two new rules in addition to the ones presented in section 1.1 for FO, namely

- $\mathfrak{A}, \theta \models \bigwedge_{i\in\mathbb{N}} \varphi_i$ whenever $\mathfrak{A}, \theta \models \varphi_i$ for all $i \in \mathbb{N}$,
- $\mathfrak{A}, \theta \models \bigvee_{i\in\mathbb{N}} \varphi_i$ whenever $\mathfrak{A}, \theta \models \varphi_i$ for some $i \in \mathbb{N}$.

The definition of semantics for conjunctions and disjunctions of greater length and homogeneous strings of quantifiers is similar. In addition to allowing infinite first-order formulas one can also allow second-order quantification and mix it with infinitary formulas. One interesting possibility is to allow one existential second-order quantifier over a countable set of second-order relations followed by an infinite conjunction of FO formulas. Such formulas, $\exists \{R_i\}_{i\in\mathbb{N}} \bigwedge_{j\in\mathbb{N}} \varphi_j$, where φ_j are FO formulas over the extended alphabet $\tau \cup \{R_i\}_{i\in\mathbb{N}}$, are called PC_Δ formulas and are intimately related to game quantification, as will be explained later.

One interesting extension of FO that is not directly included in $\mathsf{L}_{\kappa\lambda}$ is the case of an infinite string of alternating quantifiers, i.e. formulas of the form $\exists x_0 \forall x_1 \exists x_2 \forall x_3 \ldots R(x_0, x_1, \ldots)$, where $R \subseteq A^\omega$ is a relation on sequences of elements of A. The semantics of such formulas is given using two-person games known as Gale-Stewart games. The intuition behind the game is that the first player, sometimes called the Verifier, starts by choosing $x_0 \in A$. Then the second player, called the Falsifier, answers with $x_1 \in A$. After this the Verifier chooses x_2, the Falsifier answers and so on. Finally, the winner is determined depending on whether the infinite sequence that was chosen belongs to R or not. Formally, given a strategy for one player, defined as a function $f : \bigcup_{n\in\mathbb{N}} A^{2n} \to A$, and one for the other player defined as $g : \bigcup_{n\in\mathbb{N}} A^{2n+1} \to A$, we construct the sequence $\pi = \pi(f, g) = x_0 x_1 x_2 \ldots$ given by $x_{2n} = f(\pi|_{2n})$ and $x_{2n+1} = g(\pi|_{2n+1})$ for all $n \in \mathbb{N}$. This allows us to define the semantics to a formula $\varphi = \exists x_0 \forall x_1 \ldots R(\overline{x})$ by saying that φ holds in $\mathfrak{A}$ whenever there exists a strategy f for the first player such that for all counter-strategies g of the second player the play $\pi(f, g) \in R$.

In the paragraph above we did not specify the relation R, so we did not give a precise extension of FO. For this purpose, we are going to use infinitary conjunctions and disjunctions, as presented before in the context of $\mathsf{L}_{\omega_1\omega}$. Therefore we define that $\varphi(\overline{z})$ is an *open game formula* if

$$\varphi(\overline{z}) = \exists x_0 \forall y_0 \exists x_1 \forall y_1 \ldots \bigvee_{i\in\mathbb{N}} \varphi_i(\overline{z}, x_0, y_0, \ldots, x_{i-1}, y_{i-1}),$$

and $\psi(\overline{z})$ is a *closed game formula* if

$$\psi(\overline{z}) = \forall x_0 \exists y_0 \forall x_1 \exists y_1 \ldots \bigwedge_{i \in \mathbb{N}} \psi_i(\overline{z}, x_0, y_0, \ldots, x_{i-1}, y_{i-1}),$$

where φ_i, ψ_i are standard FO formulas. A structure $\mathfrak{A}$ with a mapping $\theta : \overline{z} \to \overline{a}$ is a model of $\varphi(\overline{z})$ if there exists a strategy $f : \bigcup_{n \in \mathbb{N}} A^{2n} \to A$ such that for all strategies $g : \bigcup_{n \in \mathbb{N}} A^{2n+1} \to A$ there exists $i \in \mathbb{N}$ for which

$$\varphi_i^{\mathfrak{A}}(\overline{a}, \pi(f,g)[0], \pi(f,g)[1], \ldots, \pi(f,g)[2i-1])$$

holds. Similarly, $\psi(\overline{z})$ holds if there exists a strategy $g : \bigcup_{n \in \mathbb{N}} A^{2n+1} \to A$ such that for all strategies $f : \bigcup_{n \in \mathbb{N}} A^{2n} \to A$ and for all $i \in \mathbb{N}$ the relation

$$\psi_i^{\mathfrak{A}}(\overline{a}, \pi(f,g)[0], \pi(f,g)[1], \ldots, \pi(f,g)[2i-1])$$

holds in $\mathfrak{A}$.

Observe that the definition above does not imply that a negation of an open game formula is a closed game formula, as for each of them to be true there must exist a strategy for one player that is winning against all counterstrategies. For the negation of such a formula to be true it suffices that a player can counter all strategies of the opponent with a winning strategy. The property of a two-player game that either one or the other player has a winning strategy is called *determinacy*.

It is an important result of Gale and Stewart [32] that all Gale-Stewart games with open and closed winning conditions, i.e. for relations R denoting open or closed sets in the product topology on A^ω, are determined. Since relations defined by infinite disjunctions of FO formulas are indeed open in the product topology and the ones defined by infinite conjunctions are closed, this result implies that the negation of a closed game formula is indeed an open game formula, and vice versa. This determinacy theorem, later extended by Martin [64] to all Borel winning conditions, gives a foundation for the study of game quantification.

As mentioned before, there is an intimate relationship between closed game formulas and PC_Δ formulas, which was first shown by Svenonius. For any formula $\varphi(\overline{z}) \in \mathsf{PC}_\Delta$ there exists a closed game formula $\psi(\overline{z})$ such that for all structures $\mathfrak{A}$ it holds that $\mathfrak{A} \models \varphi(\overline{z}) \to \psi(\overline{z})$ and for all *countable* structures $\mathfrak{A}$ it holds that $\mathfrak{A} \models \varphi(\overline{z}) \leftrightarrow \psi(\overline{z})$.

The above result was extended by Vaught to an analogous relationship between formulas of the form $\exists \{R_i\}_{i \in \mathbb{N}} \varphi$ where $\varphi \in \mathsf{L}_{\omega_1 \omega}$ is written using a signature extended with $\{R_i\}_{i \in \mathbb{N}}$, and *closed Vaught formulas*. These formulas are extensions of closed game formulas where countable conjunctions are allowed after each $\forall$ quantifier and countable disjunction after each $\exists$ quantifier. Their semantics is again given using a Gale-Stewart game, where the strategies additionally pick a branch of the infinite conjunction or disjunction in each step.

For any closed Vaught formula φ one can consider a finite part of the infinite sequence of alternating quantifiers in the prefix, which is an $L_{\omega_1\omega}$ formula. We say that these formulas *approximate* φ and Vaught proved that on countable structures the conjunction of all these approximating formulas is equivalent to φ. This is a strong result and combined with the Svenonius-Vaught theorem mentioned before it allows to approximate existential second-order quantification added to $L_{\omega_1\omega}$ using only pure $L_{\omega_1\omega}$ formulas. This has interesting applications to the model theory of $L_{\omega_1\omega}$, for example allowing to prove compactness and interpolation theorems for this logic. A more thorough introduction to game quantification and its applications in logic is given in [56].

2.2   Game Quantification over Infinite Words

In the closed or open game formulas $\exists x_0 \forall x_1 \dots R(\overline{x})$ the relation R is either an open or a closed set over A^ω because it is expressed by an infinite conjunction or disjunction of FO formulas. As we are interested in automatic presentations, i.e. structures over $\Sigma^{\leq\omega}$ where all relations are ω-regular, it is natural to consider ω-regular relations R instead of open or closed ones.

To extend the notion of game quantification to an automatic presentation $\mathfrak{A} = (\Sigma^{\leq\omega}, R_1, \dots, R_k)$, we make explicit use of the fact that elements of the universe are words and so already have an inductive structure to play the game on, and we introduce FO[∂], first-order logic extended with the regular game quantifier ∂. We define the meaning of the formula $\partial xy\,\varphi(x,y)$ by saying that $\partial xy\,\varphi(x,y)$ holds if φ can be satisfied by two arguments x and y which are words constructed stepwise by two opposing players. The first letter of x is given by the first player, then the first letter of y is given by the second player, then another letter of x by the first player, and so on. Formally, to capture both finite and infinite words over Σ we again define $\Sigma_\square = \Sigma \cup \{\square\}$ and set

$$\partial xy\,\varphi(x,y) \iff (\exists \text{ well-formed } f \ : \ \Sigma_\square^* \times \Sigma_\square^* \to \Sigma_\square)$$
$$(\forall \text{ well-formed } g \ : \ \Sigma_\square^* \times \Sigma_\square^* \to \Sigma_\square)$$
$$\varphi(x_{fg}, y_{fg}),$$

where x_{fg} and y_{fg} are the Σ-words constructed inductively using f and g up to the first appearance of $\square$,

$$x_{fg}[n] = f(x_{fg}|_n, y_{fg}|_n),$$

$$y_{fg}[n] = g(x_{fg}|_{n+1}, y_{fg}|_n),$$

and well-formedness means that if any of the functions f or g outputs $\square$ then the word x_{fg} resp. y_{fg} is considered to be finite and the function must then continue to output $\square$ infinitely. Formally, we say that h is well-formed when

$$h(w, u) = \square \implies (\forall w' \sqsupseteq w)\,(\forall u' \sqsupseteq u)\, h(w', u') = \square.$$

This definition coincides with the traditional one for an infinite string of alternating quantifiers over letters and a regular relation in scope of all the quantifiers,

$$\partial xy\; \varphi(x,y) \iff (\exists a_0 \forall b_0 \exists a_1 \forall b_1 \ldots)\; \varphi(a_0 a_1 \ldots, b_0 b_1 \ldots).$$

Moreover, using our notation $\partial xy\; \varphi(x)$ is equivalent to $\exists x\; \varphi(x)$ as we can always forget opponent moves and play letters from x or conversely use any g to obtain the witness x. Similarly, $\partial xy\; \varphi(y)$ is equivalent to $\forall y\; \varphi(y)$. Thus, we do not need to consider the standard quantifiers when the regular game quantifier is present.

On some structures it is possible to encode a pair of words into a single one, but that is not always the case. Therefore we might sometimes need to use the game quantifier with more variables:

$$\partial x_1 \ldots x_k y_1 \ldots y_m \; \varphi(\overline{x}, \overline{y}) \iff$$
$$(\exists f \; : \; (\Sigma_\square^*)^k \times (\Sigma_\square^*)^m \to \Sigma_\square^k)$$
$$(\forall g \; : \; (\Sigma_\square^*)^k \times (\Sigma_\square^*)^m \to \Sigma_\square^m)$$
$$\varphi(\overline{x_{fg}}, \overline{y_{fg}}),$$

where again the functions must be well–formed in each column and

$$\overline{x_{fg}}[n] = f(\overline{x_{fg}}|_n, \overline{y_{fg}}|_n), \qquad \overline{y_{fg}}[n] = g(\overline{x_{fg}}|_{n+1}, \overline{y_{fg}}|_n).$$

Example 2.1. To illustrate the use of game quantifier let us consider the following relation

$$R(u,w,s,t) \text{ defined by } \partial xy\; (y = u \to x = s) \land (y = w \to x = t).$$

We claim that R means that the common prefix of s and t is longer than the common prefix of u and w. Denoting by $v \sqcap r$ the common prefix of v and r and by $|v|$ the length of v we can say that

$$R(u,w,s,t) \equiv |u \sqcap w| < |s \sqcap t|,$$

with the additional necessary assumption that $u \neq w$ and $s \neq t$.

The intuitive way to evaluate such a formula is by means of a game played by two players – the Verifier choosing letters of x and the Falsifier choosing letters of y. To see the above equivalence, let us assume that indeed the common prefix of s and t is longer than the common prefix of u and w. In this case, the Falsifier will have to choose whether $y = u$ or $y = w$ before the Verifier chooses if $x = s$ or if $x = t$, and therefore the Verifier is going to win. In the other case, the Falsifier can win and prove the formula false as he knows if the prefix of x can be prolonged to s or to t before choosing whether $y = u$ or $y = w$.

2.3 Decidability and Determinacy for $\mathsf{FO}[\eth]$

The two basic properties of $\mathsf{FO}[\eth]$ that interest us are decidability of the model-checking problem for this logic on ω-automatic presentations and the existence of a negation normal form, which semantically corresponds to the determinacy of the underlying games.

To be able to state the existence of a negation normal form, let us introduce another variation of the regular game quantifier, namely one where it is the Falsifier who makes the moves first. Formally, let

$$\eth^\forall xy\ \varphi(x,y) \iff (\exists f : \Sigma_\square^* \times \Sigma_\square^* \to \Sigma_\square)$$
$$(\forall g : \Sigma_\square^* \times \Sigma_\square^* \to \Sigma_\square)\ \varphi(x_{fg}^\forall, y_{fg}^\forall),$$

where again the functions must be well-formed, but this time the words are constructed in reverse order,

$$y_{fg}^\forall[n] = g(x_{fg}^\forall|n, y_{fg}^\forall|n),\qquad x_{fg}^\forall[n] = f(x_{fg}^\forall|n, y_{fg}^\forall|n+1).$$

If we denote the game quantifier introduced before by $\eth^\exists$ then the intended relation that leads to negation normal form can be stated in the following way (note that the variables are reversed after $\eth^\forall$ below):

$$\eth^\exists xy\ \varphi(x,y) \equiv \neg\eth^\forall yx\ \neg\varphi(x,y).$$

When the relation of prefixing a word with a letter, written $y = ax$ for a letter $a \in \Sigma$, is present, the quantifier $\eth^\forall$ is superfluous and can be eliminated by adding one letter,

$$\eth^\forall xy\ \varphi(x,y) \iff \eth^\exists zy\ \exists x\ z = ax \wedge \varphi(x,y).$$

To verify this equivalence, observe that on the right side the Verifier must start with an a and later play a strategy that ensures that φ is satisfied, so the same strategy without the first a can be used on the left side. Conversely, if Verifier's strategy on the left side is given then playing an a and later the same strategy is winning for the right side.

The observation that we use to prove both decidability and the existence of negation normal is that if one starts with ω-regular relations then anything defined in the $\mathsf{FO}[\eth]$ logic remains ω-regular. The proof relies on the fact that, when applied to an automaton, the game quantifier indeed constructs a game and changes the automaton to an alternating one.

Lemma 2.2. *If the relation $R(\overline{x}, \overline{y}, \overline{z})$ is ω-regular over $\overline{x} \otimes \overline{y} \otimes \overline{z}$ then the relation $S(\overline{z}) \iff \eth\overline{x}\overline{y}\ R(\overline{x}, \overline{y}, \overline{z})$ is ω-regular over $\otimes\overline{z}$.*

Proof. Let us take the deterministic automaton $\mathcal{A}_R$ for R over $\overline{x} \otimes \overline{y} \otimes \overline{z}$ and construct an alternating automaton $\mathcal{A}_S$ for S over $\otimes\overline{z}$ in the following way. The set of states, acceptance condition and initial state remain the same and the new transition relation is defined by

$$\delta_S(q, \overline{c}) \;=\; \bigvee_{\overline{a} \in \Sigma_\square^k} \bigwedge_{\overline{b} \in \Sigma_\square^l} \delta_R(q, \overline{a} \otimes \overline{b} \otimes \overline{c}),$$

where k is the length of $\overline{x}$ and l is the length of $\overline{y}$.

By definition, the semantics of the relation S is

$$S(\overline{z}) \iff (\exists f \;:\; (\Sigma_\square^*)^k \times (\Sigma_\square^*)^l \to \Sigma_\square^k)$$
$$(\forall g \;:\; (\Sigma_\square^*)^k \times (\Sigma_\square^*)^l \to \Sigma_\square^l) \;\; \varphi(\overline{x}_{fg}, \overline{y}_{fg}, \overline{z}).$$

Assuming for a tuple of words $\overline{w}$ that $S(\overline{w})$ holds, we construct an accepting run ρ of the automaton $\mathcal{A}_S$ on $\overline{w}$. The run ρ is constructed inductively starting from q_0, and in parallel we construct the tuples of words $\overline{x}$ and $\overline{y}$, starting from empty words. Assuming that we are on level n in the run-tree ρ on the branch $q_0 \ldots q_{n-1}$ and that the prefixes $\overline{x}|_n$ and $\overline{y}|_n$ were constructed, we let $\overline{x}[n] = f(\overline{x}|_n, \overline{y}|_n)$ and for each $\overline{a} \in \Sigma_\square^l$ we add a branch in ρ from q_{n-1} to $q_n = \delta_R(q_{n-1}, \overline{x}[n] \otimes \overline{a} \otimes \overline{w}[n])$. Finally, we progress to one of these q_n and store $\overline{y}[n] = \overline{a}$. You can see that the function f in this definition of $S(\overline{z})$ corresponds to the choice of a branch to satisfy in the disjunction over the letters for $\overline{x}$ in the Boolean formula when selecting the run of the alternating automaton, and that the function g corresponds to the choice of the branch of the run, as all branches must be accepted. The converse direction, constructing a function f from the run ρ is analogous, the run-tree is in fact a representation of such a function. $\qquad\square$

From this lemma, together with Theorem 1.2, the decidability of FO[∂] on automatic presentations follows. The doubly exponential bound on the size of the deterministic automaton constructed from an alternating one gives a bound on the complexity of model-checking FO[∂] when the quantifier depth of a formula is fixed. It is necessary to bound the quantifier depth to get elementary complexity, as the model-checking problem on automatic structures in general is non-elementary even for FO. As regular relations on ω-words are Borel, we can derive from the previous lemma and the result of Martin [64] that the games corresponding to regular game quantifier are determined, which proves game quantifier inversion as stated below.

Corollary 2.3. FO[∂] *is decidable on ω-automatic presentations, all relations definable in it are ω-automatic and model-checking formulas with a fixed quantifier depth k is in $2k$EXPTIME.*

Corollary 2.4. *For each* FO[∂] *formula φ, each automatic presentation $\mathfrak{A}$ and each valuation θ*

$$\mathfrak{A}, \theta \models \partial^{\exists}\overline{xy}\, \varphi(\overline{x}, \overline{y}, \overline{z}) \iff \mathfrak{A}, \theta \models \neg\partial^{\forall}\overline{yx}\, \neg\varphi(\overline{x}, \overline{y}, \overline{z}).$$

2.4 Expressive Power of FO[∂]

As mentioned in the introduction, the binary ω-tree with the equal-length relation $\mathfrak{T}_{\mathsf{el}}^{\omega}(2)$ is a complete ω-automatic structure. In particular, every

ω-automatic relation over $\{0,1\}^\omega$ can be defined by an FO formula over this structure. Since FO[∂] preserves regularity by Lemma 2.2, it follows that FO[∂] is equally expressive as just FO on $\mathfrak{T}_{\mathsf{el}}^\omega(2)$.

The situation changes when the equal-length relation is not included, as already $\mathfrak{T}^\omega(2)$ is not a complete automatic structure for FO. It turns out that we can even leave out the prefix relation and still define all regular relations in FO[∂] using only the successor predicates.

Theorem 2.5. *On the structure* $(\{0,1\}^{\leq\omega}, S_0, S_1)$ *where* $S_i(v)$ *holds exactly when* $v = ui$, *all regular relations can be defined in* FO[∂].

Proof. First let us recall a few basic formulas that we are going to use. As we have already shown in Example 2.1, we can use the game quantifier to talk about the length of common prefix of words, i.e. for $u \neq w, s \neq t$ we can say $|s \sqcap t| < |u \sqcap w|$ and the other variants with $\leq, =, \geq$ and $>$ can be expressed using Boolean combinations and argument permutations of the above.

To say that x is a prefix of y we are going to say that no word $z \neq x$ has a longer common prefix with x than y,

$$x \sqsubseteq y \;\equiv\; (x = y) \vee \forall z \neq x \; |x \sqcap z| \leq |x \sqcap y|.$$

To define equal length we again use the $|s \sqcap t| < |u \sqcap w|$ relation to define that $|x| \leq |y|$. Note that so far we expressed $|s \sqcap t| < |u \sqcap w|$ only for $s \neq t$ and $u \neq w$, so we can not just write $|x \sqcap x| \leq |y \sqcap y|$. Instead, we say that for any $x' \neq x$ there is an $y' \neq y$ that has common prefix with y not shorter that the common prefix of x' and x:

$$|x| \leq |y| \;\equiv\; \forall x' \neq x \; \exists y' \neq y \; |x \sqcap x'| \leq |y \sqcap y'|.$$

Now we can use a Boolean combination to define $|x| = |y|$, and in this way we obtain both $\sqsubseteq$ and el. As all ω-regular relations over $\mathfrak{T}_{\mathsf{el}}^\omega(2)$ can be defined in FO, this completes the proof. $\square$

Note that the above formulas for $\sqsubseteq$ and el did not involve the successor predicates. Definability of these two relations on any automatic presentation in FO[∂] just with equality is an important property which will be used subsequently.

One interesting example of an automatic presentation is the binary coding of natural numbers where the least significant digit comes first. We look at the expressive power of FO[∂] on such presentations over $\{0,1\}^{\leq\omega}$. To speak meaningfully about numbers, as opposed to words representing them, there must be a relation eq in such presentation that defines the equality between numbers as opposed to equality over words with redundant zeros, $\mathsf{eq}(x,y) \equiv (x = n0^k$ and $y = n0^l)$ for some $k, l \in \mathbb{N}$. Using eq and $\sqsubseteq$ it is possible to define S_0 and S_1 over $\{0,1\}^{\leq\omega}$ as words ending with one are exactly those without redundant zeros. Thus in any such presentation with eq it is again possible to define all regular relations in FO[∂]. This can as well be used to define $+$ and thus adding other strong non-regular relations to the structure, for example multiplication, makes model-checking undecidable.

Corollary 2.6. *On the binary (lower-endian) presentation of $(\mathbb{N}, =)$ the relations $+$ and $|_2$ (and all relations regular in this presentation) are definable in* $\mathsf{FO}[\partial]$. *On the binary presentation of Skolem arithmetic $(\mathbb{N}, =, \cdot)$ the logic* $\mathsf{FO}[\partial]$ *is undecidable.*

2.5 Inductive Automorphisms

After analyzing what can be expressed in $\mathsf{FO}[\partial]$, we want to look for methods of establishing which relations can *not* be expressed in this logic. For example, one could ask whether a^ω can be expressed in $\mathsf{FO}[\partial]$ without any relations other than equality of words on $\{a, b\}^{\leq\omega}$. We are going to develop a general method to answer such questions by showing that there is a class of automorphisms of a structure that extend to all relations definable in $\mathsf{FO}[\partial]$.

First of all, observe that not all automorphisms of an automatic presentation, when considered just as a first-order structure $(\Sigma^{\leq\omega}, R_1, \ldots, R_k)$, extend to relations definable in $\mathsf{FO}[\partial]$. For example, on a presentation with no relations, the bijection of $\Sigma^{\leq\omega}$ that swaps a^ω with b^ω and leaves other elements untouched is an automorphism. On the other hand, the relation $|s \sqcap t| < |u \sqcap w|$ is definable in $\mathsf{FO}[\partial]$ just with equality, but the bijection described above is not an automorphism of the structure extended with this relation, since

$$|b^\omega \sqcap ab^\omega| < |a^\omega \sqcap ab^\omega| \quad \text{but} \quad |a^\omega \sqcap ab^\omega| > |b^\omega \sqcap ab^\omega|.$$

The example above is not surprising since we proved that the prefix relation is definable in $\mathsf{FO}[\partial]$ on any presentation, so any automorphism that preserves $\mathsf{FO}[\partial]$-definable relations on $(\Sigma^{\leq\omega}, R_1, \ldots, R_k)$ must be an automorphism of $(\Sigma^{\leq\omega}, R_1, \ldots, R_k, \sqsubseteq)$. We are going to explicitly define the class of *inductive automorphisms* that do extend to relations definable in $\mathsf{FO}[\partial]$ by restricting the bijections of $\Sigma^{\leq\omega}$ to a special form. It turns out that this class is precisely the class of all automorphisms of $(\Sigma^{\leq\omega}, \sqsubseteq)$, so extending the presentation with the prefix relation assures that all $\mathsf{FO}[\partial]$-definable relations are preserved under automorphism.

Definition 2.7. *The bijection* $\phi : \Sigma^{\leq\omega} \to \Sigma^{\leq\omega}$ *is* inductive *whenever it does not change the length of the words, i.e. $|\phi(u)| = |u|$ for every word u, and additionally there exists a family of permutations*

$$\{\pi_w\}_{w \in \Sigma^*} \quad \pi_w : \Sigma \to \Sigma,$$

such that for each word u with at least n letters the nth letter of $\pi(u)$ is given by the appropriate permutation,

$$\phi(u)[n] = \pi_{u|_{n-1}}(u[n]).$$

Observe that the inverse bijection ϕ^{-1} of any inductive bijection ϕ is again inductive as inverse permutations $\{\pi_w^{-1}\}$ can be used.

If we restrict our attention to an automorphism ϕ that is an inductive bijection then the structure can be extended with any $\mathsf{FO}[\partial]$ definable relation and ϕ will still be an automorphism of the extended structure, as formulated below.

Theorem 2.8. *Let ϕ be an inductive automorphism of a structure $\mathfrak{A} = (\Sigma^{\leq\omega}, R_1, \ldots, R_k)$ and $R(\overline{x})$ a relation defined by an $\mathsf{FO}[\partial]$ formula $\varphi(\overline{x})$. Then ϕ is an automorphism of the extended structure $(\Sigma^{\leq\omega}, R_1, \ldots, R_k, R)$.*

Proof. We proceed by induction on the structure of formulas and it is enough to consider the inductive step for the game quantifier. Let $\varphi(\overline{x}, \overline{y}, \overline{z})$ be a formula such that

$$\varphi^{\mathfrak{A}}(\overline{a}, \overline{b}, \overline{c}) \iff \varphi^{\mathfrak{A}}(\phi(\overline{a}), \phi(\overline{b}), \phi(\overline{c})).$$

We show that for $\psi(\overline{z}) = \partial\overline{x}\,\overline{y}\,\varphi(\overline{x}, \overline{y}, \overline{z})$ it holds $\psi^{\mathfrak{A}}(\overline{c}) \iff \psi^{\mathfrak{A}}(\phi(\overline{c}))$.

To prove it let us define for any strategies f of the Verifier and g of the Falsifier used in $\partial\overline{x}\,\overline{y}\,\varphi(\overline{x}, \overline{y}, \overline{z})$ the transposed strategies f_ϕ, g_ϕ in the following way:

$$f_\phi(\overline{u}, \overline{w}) = \pi_{\phi^{-1}(\overline{u})}f(\phi^{-1}(\overline{u}), \phi^{-1}(\overline{w})),$$

$$g_\phi(\overline{u}, \overline{w}) = \pi_{\phi^{-1}(\overline{w})}g(\phi^{-1}(\overline{u}), \phi^{-1}(\overline{w})),$$

where π_w is the permutation for word w associated with ϕ and $\pi_{\overline{w}}$ applied to a tuple $\overline{v}$ of the same length means applying π_{w_i} to each element v_i. You should observe that when the players play with strategies f_ϕ, g_ϕ then the resulting words are exactly images of the words that result from using f and g under ϕ,

$$\overline{x_{f_\phi g_\phi}} = \phi(\overline{x_{fg}}), \qquad \overline{y_{f_\phi g_\phi}} = \phi(\overline{y_{fg}}).$$

In this way we can use the winning strategy f for the first player in $\psi(\overline{z})$ and play with f_ϕ in $\psi(\phi(\overline{z}))$. If the opponent chooses to play g then in the end the formula $\varphi(\overline{x_{f_\phi g}}, \overline{y_{f_\phi g}}, \phi(\overline{z}))$ will be evaluated, but

$$\varphi(\overline{x_{f_\phi g}}, \overline{y_{f_\phi g}}, \phi(\overline{z})) \equiv \varphi(\phi(\overline{x_{fg_{\phi^{-1}}}}), \phi(\overline{y_{fg_{\phi^{-1}}}}), \phi(\overline{z}))$$

$$\equiv \varphi(\overline{x_{fg_{\phi^{-1}}}}, \overline{y_{fg_{\phi^{-1}}}}, \overline{z}),$$

which holds as f is winning against any strategy, in particular against $g_{\phi^{-1}}$. $\square$

While the explicit definition of inductive automorphisms given above was useful for the proof, we can characterize these automorphisms in another way, namely as automorphisms of $(\Sigma^{\leq\omega}, \sqsubseteq)$. On the one hand, any inductive automorphism preserves $\sqsubseteq$ because this is an $\mathsf{FO}[\partial]$-definable relation. On the other hand, it can be shown by induction on the prefix order that any automorphism of $(\Sigma^{\leq\omega}, \sqsubseteq)$ is inductive. First, the empty word is preserved by any automorphism that preserves the prefix relation as it is the minimal element of $\sqsubseteq$. Secondly, if a word w is mapped to w' by an automorphism preserving $\sqsubseteq$, then all $\sqsubseteq$-successors of w must be mapped to $\sqsubseteq$-successors of w', which defines the permutation π_w. Thus, all automorphisms of an automatic presentation $(\Sigma^{\leq\omega}, \sqsubseteq, R_1, \ldots, R_k)$ preserve $\mathsf{FO}[\partial]$-definable relations.

The standard way to show that a relation is not definable in a logic using automorphisms is to find an automorphism the relation is not invariant under. Theorem 2.8 makes it possible to use this standard method for $\mathsf{FO}[\partial]$ on automatic presentations, as shown in the following example, where we answer the question asked at the beginning of this section.

Example 2.9. Let us consider the automorphism ϕ of $\{a, b\}^{\leq\omega}$ that just swaps the first letter of all words, i.e. $\phi(au) = bu$, $\phi(bv) = av$, $\phi(\varepsilon) = \varepsilon$. The mapping ϕ is an inductive bijection; the appropriate permutations π_w are identities for all $w \neq \varepsilon$, and π_ε is given by $\pi_\varepsilon(a) = b$ and $\pi_\varepsilon(b) = a$. This automorphism maps a^ω to ba^ω and thus the set $\{a^\omega\}$ is not preserved under ϕ. By Theorem 2.8 we conclude that a^ω is not definable in $\mathsf{FO}[\partial]$ just with equality.

3

Games for Model Checking on Automatic Structures

In the previous chapter we used games as a tool to define the semantics of game quantification and to investigate questions in logic. In this chapter we focus on games in their own right.

We start by defining games played on graphs by two players with perfect information. The connection between such games and logic is illustrated on two well-known examples: the game-theoretical semantics of first-order logic where games of finite duration are used, and model-checking of modal μ-calculus where parity games are appropriate. These two examples show that studying the relation to games can both lead to better insight into the expressive power of a logic and also have an algorithmic utility for model checking. This motivates us to look for games for model-checking on automatic structures.

To find an appropriate game model for first-order logic on an automatic structure, we fix a presentation of the structure and investigate the extended logic FO[∂]. For this setting, we introduce multiplayer games played by two coalitions with opposing objectives and with imperfect information exchanged according to a hierarchical constraint [50]. On the one hand, this constraint is suitable for defining model-checking games for the extended first-order logic, and it is necessary for the problem of establishing the winning coalition to be decidable. On the other hand, this constraint alone is not sufficient for establishing the winners to be decidable.

To identify the properties needed to make hierarchical games decidable, we study a restricted version of these games where players are forced to alternate. We show that this constraint is required both for determinacy of hierarchical games and for decidability of the problem of establishing the winning coalition. Finally, we prove that hierarchical games where players alternate are indeed model-checking games for FO[∂] on automatic presentations.

3.1 Games on Graphs and Logic

In the previous chapter we discussed game quantification and used Gale-Stewart games to provide semantics for game formulas. Since we were only interested in the existence of winning strategies, we did not give a formal definition of what a game is in that context. In this section we want to take a step back and define games, more precisely games played on graphs. We also give an overview of the well-known connection between two-player zero-sum games with complete information and fixed-point logics.

The intuition behind a two-player zero-sum turn-based game played on a graph is very natural. Two players, let us call them Player 0 and Player 1, play by moving a token around a graph of positions. There is a position singled out in which the game starts and every position is assigned to one of the players. When the token is in a position that belongs to one of the players, this player is required to move by choosing an edge going out from this position. If there are no outgoing edges, the player who can not move loses. If the players manage to keep playing infinitely long, then the winner is decided based on a winning condition that specifies which infinite plays are winning for Player 0 and which for Player 1.

Definition 3.1. *A Büchi, parity, Streett, Rabin or Muller game is given by a tuple $\mathcal{G} = (V_0, V_1, E, \mathcal{F})$ where V_0 is the set of positions of Player 0 and V_1, disjoint from V_0, contains the positions of Player 1. $E \subseteq V \times V$ is the edge relation denoting possible moves between positions $V = V_0 \cup V_1$, and $\mathcal{F} \subseteq V^\omega$ is a winning condition, represented in the same way as Büchi, parity, Streett, Rabin and Muller acceptance conditions for automata described in section 1.3.*

To avoid tedious case distinctions, we often assume that all plays are infinite, i.e. that $vE \neq \emptyset$ for all $v \in V$.

You can see that the Gale-Stewart game for a structure $\mathfrak{A}$ can be viewed as a graph game, either as a game on the tree $\mathfrak{T}(A)$ with players alternating their moves or as a game on the complete bipartite graph $A \times A$ with one side belonging to Player 0 and the other to Player 1.

A strategy for player $i \in \{0, 1\}$ in the game $\mathcal{G}$ is a function $\sigma : V^* V_i \to V$ with $(v, \sigma(hv)) \in E$ for all $h \in V^*$ and $v \in V_i$. A play $\pi = v_0 v_1 \ldots$ is *consistent with a strategy* σ for player i if $v_{n+1} = \sigma(v_0 \ldots v_n)$ for every n such that $v_n \in V_i$. Given strategies σ, ρ for Player 0 and Player 1, respectively, we denote by $\pi_{\sigma,\rho}(v_0)$ the unique play starting in position v_0 which is consistent with both σ and ρ.

We say that a strategy σ is winning for Player 0 from v_0 if for all strategies ρ of the opponent $\pi_{\sigma,\rho}(v_0) \in \mathcal{F}$. Analogously, a strategy ρ is winning for Player 1 from v_0 if for all strategies σ of the opponent $\pi_{\sigma,\rho}(v_0) \notin \mathcal{F}$. The set of all positions from which player i has a winning strategy is called the winning region of player i. A game $\mathcal{G}$ is determined if from every position either Player 0 or Player 1 has a winning strategy. Thus, in a determined game, the game graph can be partitioned into winning regions of Player 0 and Player 1.

In many cases one is interested not only in arbitrary winning strategies, but in strategies of a special kind. One prominent example are *positional strategies*, where the strategy depends only on the current position and not on the previous positions of the play, i.e. $\sigma(hv) = \sigma(v)$ for any history h. In a stronger version of determinacy one requires the winning strategies to belong to a certain class. For example, games with parity winning conditions are determined in positional strategies [31, 68], i.e. from every position either Player 0 or Player 1 has a positional strategy that is winning against all strategies of the opponent. For games with Muller winning conditions on finitely many priorities a larger class of strategies is needed, namely such where a finite number of memory states is allowed. We investigate various kinds of determinacy and memory for strategies in chapter 4.

There is an intimate connection between zero-sum games and logic. The idea to give semantics to logics using games was mentioned already in the last decade of the 19th century by C.S. Pierce, and about sixty years later Paul Lorenzen gave a game-theoretical semantics for first-order logic. Giving a game-theoretical semantics to a logic means that for the evaluation of a formula φ on a structure $\mathfrak{A}$ one constructs a model-checking game $\mathsf{MC}(\mathfrak{A}, \varphi)$ such that Player 0 has a winning strategy in $\mathsf{MC}(\mathfrak{A}, \varphi)$ from an initial position exactly if $\mathfrak{A} \models \varphi$.

The model-checking game for an FO formula φ on $\mathfrak{A}$ is constructed in a very intuitive way. The positions of the game consist of subformulas of φ together with a valuation of all free variables in the subformula. If the position is of the form $(\varphi_1 \vee \varphi_2, \theta)$ then Player 0 moves either to (φ_1, θ) or to (φ_2, θ). Analogously, from $(\varphi_1 \wedge \varphi_2, \theta)$ Player 1 moves to one of the subformulas. In a position of the form $(\exists x \varphi, \theta)$, Player 0 moves by choosing an element $a \in \mathfrak{A}$. The next position is then $(\varphi, \theta[x \leftarrow a])$. For $(\forall x \varphi, \theta)$, the other player can make analogous moves. When the game reaches a position (φ, θ) for an atomic formula φ, the winner is determined depending on whether or not $\mathfrak{A}, \theta \models \varphi$.

On finite structures first-order logic is often too weak to express properties of interest. Before we proceed to show model-checking games for first-order logic on infinite structures, let us recall how a more expressive logic, the modal fixed-point logic, can be model-checked on finite structures using parity games.

In computer science, real-world systems are often modeled using finite Kripke structures, which are directed graphs labeled by a set of predicates. Formally, a Kripke structure is a tuple $\mathcal{K} = (V, E, P_1, \ldots, P_k)$ with $E \subseteq V \times V$ and $P_i \subseteq V$. Important properties that often need to be checked on such systems include *reachability*, i.e. the question whether a node where a predicate P_i holds can be reached from an initial node, and *safety*, i.e. the question whether nodes where a predicate P_j holds can be avoided on all possible paths from an initial node. These properties are not definable in FO, but there are well-known temporal logics, like the linear time logic LTL and the branching-time logic CTL, which can express these properties. There is an elegant modal logic that subsumes all these temporal logics and can express many interesting properties, the modal μ-calculus L_μ. Formulas φ of L_μ are formed according to the following syntax,

$$\varphi \ = \ P_i \mid \neg P_i \mid X \mid \varphi \wedge \varphi \mid \varphi \vee \varphi \mid \Box \varphi \mid \Diamond \varphi \mid \mu X \varphi \mid \nu X \varphi,$$

and evaluated on a Kripke structure $\mathcal{K}$ using the following semantics.

- $\mathcal{K}, v \models P_i$ whenever $P_i(v)$ holds and $\mathcal{K}, v \models \neg P_i$ in the other case,
- $\mathcal{K}, v \models \varphi \wedge \psi$ $(\varphi \vee \psi)$ whenever $\mathcal{K}, v \models \varphi$ and (or) $\mathcal{K}, v \models \psi$,
- $\mathcal{K}, v \models \Diamond\varphi$ whenever there is a $w \in vE$ for which $\mathcal{K}, w \models \varphi$ holds,
- $\mathcal{K}, v \models \Box\varphi$ whenever $\mathcal{K}, w \models \varphi$ for all $w \in vE$,
- $\mathcal{K}, v \models \mu X\varphi$ whenever $(\mathcal{K}, X), v \models \varphi$, where X is the *smallest* subset of V for which the equation $X = \{w : (\mathcal{K}, X), w \models \varphi\}$ holds,
- $\mathcal{K}, v \models \nu X\varphi$ whenever $(\mathcal{K}, X), v \models \varphi$, where X is the *biggest* subset of V for which the equation $X = \{w : (\mathcal{K}, X), w \models \varphi\}$ holds.

Note that in the syntax we use X to denote a set variable, while in the definition of semantics we write $(\mathcal{K}, X)$ for the Kripke structure $\mathcal{K}$ extended with the predicate X. The semantics above is well defined only if the smallest and biggest solutions to the fixed-point equation exist, but this is indeed the case due to the monotonicity of all the operators of L_μ.

The modal μ-calculus is a very expressive logic, in fact it can express all MSO-definable properties that are invariant under bisimulation [46], and most properties of practical interest belong to this class. To define a model-checking game $\mathsf{MC}(\mathcal{K}, \varphi)$ for an L_μ formula φ on a Kripke structure $\mathcal{K}$ one proceeds in an analogous way to first-order logic. Player 0 chooses a successor for $\Diamond$ and $\vee$, while Player 1 moves for $\Box$ and $\wedge$. Additionally, to handle fixed-point operators, from any set variable X a new edge is added back to the formula $\mu X\varphi$ or $\nu X\varphi$ where the variable X was introduced. These back-edges make infinite plays possible and it turns out that the winner of such an infinite play is decided depending on whether the outermost fixed-point variable occurring infinitely often in the play is introduced in a μ or in a ν formula. This corresponds exactly to the parity condition and indeed, not only are parity games powerful enough for model-checking L_μ, the converse holds as well, i.e. winning in any parity game (with a fixed number of priorities) can be expressed in the μ-calculus.

The correspondence between L_μ and parity games is not only an interesting extension of the analogous relation between first-order logic and games of finite duration, it also has interesting algorithmic consequences. While it is still open whether there exists a polynomial-time algorithm for model-checking L_μ, all of the most efficient algorithms known so far [47, 48, 86, 49] rely on the representation of the problem as a game. In particular, one very efficient algorithm [86] heavily exploits the structure of the game. This algorithm does not compute the fixed-points in an iterative way, as suggested by the structure of the L_μ formula. Instead, it starts by guessing a positional strategy in the parity game and then it improves this strategy, which often takes fewer steps than the iterative fixed-point evaluation. The fact that the structure of the game can be of algorithmic use is an additional motivation to look for model-checking games for $\mathsf{FO}[\partial]$ on automatic structures.

3.2 Games with Hierarchical Imperfect Information

Our goal in this section is to describe a class of games that will later be used for model-checking first-order logic with the game quantifier on presentations of automatic structures. To define such games we go beyond two-player perfect information games and use multiplayer games with imperfect information. Even though there are multiple players, in the games we define they form two coalitions with strictly opposing objectives. For this reason one could use a different metaphor with just two players for the same class of games. We use the multiplayer setting in this chapter and discuss the other possibilities in the final chapter.

While imperfect information is a standard element of classical game theory, especially for games in extensive form, in computer science games with imperfect information played on graphs have first been studied in the context of alternating Turing machines with private states [76, 77]. At that time only the reachability condition was considered. Algorithmic solutions for imperfect information games with ω-regular winning conditions were presented only recently [21], however only for the case of observable winning conditions.

The standard way to represent imperfect information in games is by means of *information sets*, equivalence relations describing which states can not be distinguished by a given player. We find it more convenient to use a different representation, in which players see some of the actions of their opponents and other actions are hidden. It is possible to transform between these two representations, but the transformation may increase the size of the game.

Definition 3.2. *A hierarchical Büchi, parity, Rabin, Streett or Muller game with actions in a finite set Σ is given by a tuple*

$$(V_{1,\mathrm{I}}, \ldots, V_{N,\mathrm{I}}, V_{1,\mathrm{II}}, \ldots, V_{N,\mathrm{II}},\ \mu,\ \mathcal{F}).$$

The game is played by two coalitions, I *and* II, *each consisting of N players, with the set of players denoted*

$$\Pi = (1, \mathrm{I}), (2, \mathrm{I}), \ldots, (N, \mathrm{I}), (1, \mathrm{II}), (2, \mathrm{II}), \ldots, (N, \mathrm{II})$$

and the arena of the game given by the pairwise disjoint sets of positions of each player, $V_{1,\mathrm{I}}, \ldots, V_{N,\mathrm{I}}, V_{1,\mathrm{II}}, \ldots, V_{N,\mathrm{II}}$. Positions of coalition I *are denoted by $V_{\mathrm{I}} = V_{1,\mathrm{I}} \cup \ldots \cup V_{N,\mathrm{I}}$ and the ones of coalition* II *by $V_{\mathrm{II}} = V_{1,\mathrm{II}} \cup \ldots \cup V_{N,\mathrm{II}}$, with all positions denoted $V = V_{\mathrm{I}} \cup V_{\mathrm{II}}$. The function $\mu : V \times \Sigma \to V$ defines the possible moves, so that when a player chooses an action $a \in \Sigma$ in his position v then the token is moved and the play proceeds to position $\mu(v, a)$. The objective of coalition* I *is given by the winning condition $\mathcal{F} \subseteq V^{\omega}$, represented in a finite way as a parity, Streett, Rabin or Muller condition, depending on the type of the game.*

When a hierarchical game is played infinitely long, an infinite sequence of actions is taken by the players during the play, which we call the *play actions* sequence and denote by $\alpha \in \Sigma^\omega$. Conversely, with every play actions sequence α and a starting position v_0, we associate the unique play $\pi_\alpha(v_0)$. It is the infinite sequence of positions that results from making the moves according to α,

$$\pi_\alpha(v_0) = v_0 v_1 \dots \quad \Longleftrightarrow \quad v_{i+1} = \mu(v_i, \alpha[i]) \text{ for all } i \in \mathbb{N}.$$

During the play $\pi_\alpha(v_0)$ we encounter a sequence of players that take the moves in each step, defined by $\Pi_\alpha(v_0)[i] = p \Leftrightarrow \pi_\alpha(v_0)[i] \in V_p$.

In a hierarchical game each player p has to decide on a strategy σ_p : $\Sigma^* \to \Sigma$. In a game with perfect information one says that play actions α are consistent with a strategy σ_p in a play starting in v_0 if for each move i taken by player p the action taken is given by the strategy acting on the history of actions, $\alpha[i] = \sigma_p(\alpha|_i)$.

Since the players do not have perfect information, we additionally assume that for each player p there is a *view function* ν_p that extracts the information visible for this player from the history of play actions. More precisely, let $\nu_p : (\Sigma \times \Pi)^* \to \Sigma^*$ be the function that extracts the information visible to player p from the history of play actions labeled by players who took these actions. We say that a sequence of play actions α is consistent with a strategy σ_p of player p in a play starting in v_0 if, for each i for which $\pi_\alpha[i] \in V_p$, it holds that

$$\alpha[i+1] = \sigma_p(\nu_p((\alpha[0], \Pi_\alpha[0]) \dots (\alpha[i], \Pi_\alpha[i]))).$$

The above definition of views of play history is very general, but we will only use a concrete special case of *hierarchical* view functions. These hierarchical views allow player k in each coalition to see the moves of players $1, \dots, k$ in both coalitions, but do not allow him to see the moves of players with numbers $j > k$. Formally, for a player $p = (k, c)$, i.e. player number k in coalition c,

$$\nu_p((a_0, p_0)(a_1, p_1) \dots (a_n, p_n)) = a_{i_1} a_{i_2} \dots a_{i_l}$$

if for all $i \in \{i_1, \dots, i_l\}$ the player $p_i = (l, d)$ has number $l \leq k$, and for all other $j \notin \{i_1, \dots, i_l\}$ the player $p_j = (m, e)$ has number $m > k$.

There is a good reason to use hierarchical view functions, namely that for most other kinds of information flow, determining the winner, even in a reachability game with three players, is undecidable [4, 2].

To define when coalition I wins a hierarchical game we can not require from all players in this coalition to put forth their winning strategies before players in coalition II do, as it is often done in games with perfect information. Intuitively, in that case players with higher numbers would lose their advantage of information as their strategies would be disclosed too early. Therefore, we use the following definition that requires that strategies are given stepwise, level by level in the information hierarchy.

Definition 3.3. *Coalition* I *wins the hierarchical game*

$$(V_{1,\mathrm{I}}, \ldots, V_{N,\mathrm{I}}, V_{1,\mathrm{II}}, \ldots, V_{N,\mathrm{II}}, \ \mu, \ \mathcal{F})$$

starting from position v_0 if the following condition holds. There exists a strategy $\sigma_{1,\mathrm{I}}$ for player 1, I, such that for each strategy $\sigma_{1,\mathrm{II}}$ of player 1, II, there exists a strategy $\sigma_{2,\mathrm{I}}$, such that for each strategy $\sigma_{2,\mathrm{II}}$, ..., there exists a strategy $\sigma_{N,\mathrm{I}}$, such that for each strategy $\sigma_{N,\mathrm{II}}$, the play actions sequence α that starts from v_0 and is consistent with all strategies $\sigma_{1,\mathrm{I}}, \sigma_{1,\mathrm{II}}, \ldots, \sigma_{N,\mathrm{I}}, \sigma_{N,\mathrm{II}}$ results in a play winning for I, i.e. $\pi_\alpha(v_0) \in \mathcal{F}$.

The definition for coalition II is analogous, i.e. there exists a $\sigma_{1,\mathrm{II}}$, such that for all $\sigma_{1,\mathrm{I}}$, ..., the play is winning for II, i.e. $\pi_\alpha(v_0) \notin \mathcal{F}$.

Example 3.4. To get an intuition about the kind of interactions that appear in hierarchical games, let us consider the simple game depicted in Figure 3.1 in two variants. The positions of coalition I are round, the positions of coalition II are square, there are two levels of information, and the positions on the upper level are dotted.

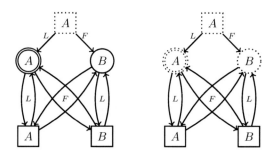

Fig. 3.1. Example of a hierarchical game in two variants

You can think of this game as played using a coin with two sides, A and B. Each of the players can choose to either flip the coin (F) or leave it as it is (L). Formally, there are four players in this game, two in each coalition. The top position belongs to 2, II and the two bottom positions belong to 1, II. The game proceeds as follows: first the second player of coalition II chooses either to flip the coin or to leave it intact. Afterward, only the other two players play by either flipping the coin or leaving it as it is. Coalition I wins if the A side of the coin is seen infinitely often in positions where players in coalition I move, as marked in Figure 3.1.

To illustrate the importance of hierarchical information levels we consider two variants of this game. In the first one (left), the bottom strongly connected component belongs to players on the same information level, i.e. to 1, II and 1, I. In this scenario, coalition II can win, because first player $2, \mathrm{II}$ can flip the coin to B and later player $1, \mathrm{II}$ can always repeat the last move of player $1, \mathrm{I}$.

In the other variant (right), the player in coalition I has more information, i.e. the bottom strongly connected component belongs to $1, \mathrm{II}$ and $2, \mathrm{I}$, with $V_{1,\mathrm{I}} = \emptyset$. In this case coalition I can win, because the strategy of player $2, \mathrm{I}$ is given after the strategy of $1, \mathrm{II}$ is set. Therefore, player $2, \mathrm{I}$ can assure that the coin will be flipped after each two moves, which guarantees that I holds the coin on the A side infinitely often, independent of the first move of $2, \mathrm{II}$.

3.3   Alternation of Moves in Hierarchical Games

In games with perfect information it is not necessary to assume that the players move in any fixed order. Moreover, the assumption that players move in an alternating way can be made without loss of generality. We show that this is not the case for hierarchical games. Thus, we define an *alternating hierarchical game* as a hierarchical game, where for each letter $a \in \Sigma$ and each level $i = 1, \ldots, N$ the following alternation conditions hold:

$$v_i \in V_{i,\mathrm{I}} \implies \mu(v_i, a) \in V_{i,\mathrm{II}},$$
$$v_i \in V_{i,\mathrm{II}} \implies \mu(v_i, a) \in V_{(i \bmod N)+1,\mathrm{I}}.$$

To see that non-alternating hierarchical games can not be reduced to alternating ones, let us consider the game depicted in Figure 3.2. The leftmost and the rightmost bottom position is winning for coalition I, while in the other two bottom positions coalition I loses. This simple hierarchical game is not alternating and we show that it is not determined. To win this game, the player on the lower level of information, i.e. $1, \mathrm{I}$ or $1, \mathrm{II}$, has to predict the move of the opponent, i.e. $1, \mathrm{II}$ or $1, \mathrm{I}$. In particular, his strategy has to start with an a exactly if the opponent starts with an a. As this holds for players in both coalitions, it leads to a non-determined game as each player can counter the strategy of the opponent, once it is known.

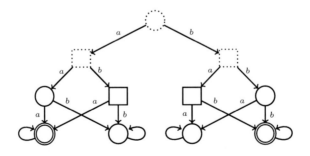

Fig. 3.2. Non-determined hierarchical game

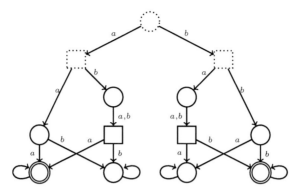

Fig. 3.3. Alternation makes hierarchical games determined

Introducing alternation of moves, even in the simplest possible way, changes this situation. The game depicted in Figure 3.3 is identical to the one in Figure 3.2 except for two additional positions of player 1, I. These positions may seem useless as there is no choice to be made there, but the new game is determined. To convince yourself that, in the extended game, coalition II can indeed win, take the following strategy of player 1, II: let him always play the opposite move to the one that was taken before by player 1, I. For player 2, II take the following strategy: if player 1, I declared that he will play a first, then play b, and else play a first. You can check that these strategies are indeed winning for coalition II, but this is possible only because when constructing the strategy for 1, II the first letter played by 1, I was already known.

Another important difference between alternating and non-alternating hierarchical games is decidability of the problem of establishing whether coalition I wins the game. We show in the next section that this problem is decidable for alternating hierarchical games, and here we prove that in the general non-alternating case it is undecidable. The differences between alternating and non-alternating hierarchical games can be explained on the level of logic and model-checking, as alternating hierarchical games correspond to model-checking on automatic presentations, while non-alternating games correspond to model-checking on presentations that use asynchronous automata, known as rational structures, which have undecidable first-order theory. It is also interesting to observe that the proof of undecidability uses the fact that all players in hierarchical games as we defined them choose actions from the same alphabet Σ. If we assume that in a hierarchical game every player chooses actions from his own alphabet, which does not overlap with the alphabet of any other player, then establishing which coalition wins is decidable even for non-alternating games, cf. [73].

Theorem 3.5. *The question whether coalition* I *wins in a hierarchical Büchi game is undecidable.*

Proof. We reduce the Post correspondence problem for $\bar{u} = u_1, \ldots, u_K$ and $\bar{v} = v_1, \ldots, v_K$, where $u_i, v_i \in \{a, b\}^*$, to the problem whether coalition I wins in the hierarchical game $\mathcal{G}_{\bar{u},\bar{v}}$. The possible actions in $\mathcal{G}_{\bar{u},\bar{v}}$ are $\Sigma = \{a, b, \square, 1, 2, \ldots, K\}$ and they intuitively correspond to the players choosing letters of the words u_i, v_i, a special delimiter $\square$, and choosing which word to play next.

In constructing $\mathcal{G}_{\bar{u},\bar{v}}$, we are going to use subgames such that, for a given word u, the subgame enforces that u is played, or else the player that moves loses. Such a subgame has one more position than the length of u, and if the wrong letter is chosen then the move leads to a position where the player loses. There is only one outgoing edge in such a subgame, the one taken when the last letter of u is played. In Figure 3.4 we depicted an example subgame for $u = aba$ and player $1, I$, who loses in the rightmost position.

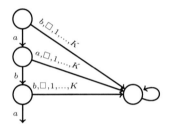

Fig. 3.4. Example subgame for $u = aba$

We start the construction of the game $\mathcal{G}_{\bar{u},\bar{v}}$ with a position belonging to player $3, II$ with two possible (non-losing) moves. In this position, coalition II can decide if the test will be done for the words $\bar{u}$ or for the words $\bar{v}$. All other positions will be on lower levels of information and we construct them in such a way that coalition I will never be able to deduce in which component the play is taking place.

Each of the two components, for $\bar{u}$ and for $\bar{v}$, starts with a position of player $2, I$ where this player chooses if he wants to play a word with index $1, \ldots, K$ or the special symbol $\square$. If the special symbol is chosen, player $1, I$ must play the same symbol $\square$ and the play returns back to the position, where $2, I$ chooses a word. When an index L is chosen, then in each of the components first the word v_L and then the word u_L is played. The difference is that, in the first component (for $\bar{u}$), it is player $2, II$ who must play v_L and player $1, I$ must play u_L, while in the other component (for $\bar{v}$), it is player $1, I$ who must play v_L and player $2, II$ who must play u_L. After the two words were played, the play returns to the position where $2, I$ chooses the index of a word to be played. The complete game is depicted in Figure 3.5, using subgames for u_i and v_i.

The winning condition is defined as follows: the special symbol $\square$ must be chosen by $2, I$ infinitely often and additionally there must be another action L, different from $\square$, that is played infinitely often. While this is not directly a

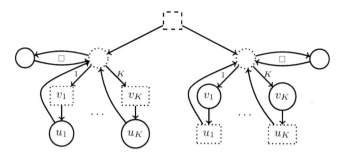

Fig. 3.5. Complete game $\mathcal{G}_{\overline{u},\overline{v}}$

Büchi condition, the game can be transformed into a game with Büchi winning condition. In the modified game, one more position for player 2, I is added in each component, with the same moves as in the original one except for the possibility of choosing $\square$. In the transformed game, when 1, I chooses $\square$ in the only position where he is allowed to do so, the play proceeds from the new position of 2, I where $\square$ is not allowed, thus ensuring that a non-$\square$ action is taken.

Let us first show that if there is a solution for the Post correspondence problem for $\overline{u}$ and $\overline{v}$ then coalition I has winning strategies for $\mathcal{G}_{\overline{u},\overline{v}}$. Indeed, let $i_1, i_2, \ldots, i_M$ be the indices for the solution of the correspondence problem, so that $u_{i_1} u_{i_2} \ldots u_{i_M} = v_{i_1} v_{i_2} \ldots v_{i_M}$. Let player 2, I choose i_1 in his first move, then i_2, i_3, and so on up to i_M, then the special symbol $\square$, and then again i_1, i_2, and so on. Player 1, I is going to play the letters from the word $u_{i_1} u_{i_2} \ldots u_{i_M}$ in turn, and then $\square$, and then again the letters $u_{i_1} u_{i_2} \ldots u_{i_M}$, and $\square$, and so on. Clearly, player 2, I chooses $\square$ and non-$\square$ infinitely often, so to show that coalition I wins we only need to prove that player 1, I will never play the wrong letter in a subgame for some word w. If the play is taking place in the $\overline{u}$ component this is clear from the definition of the strategies given above, as player 1, I plays exactly the words indices of which player 2, I chooses. When the play takes place in the $\overline{v}$ component, the indices chosen by player 2, I force player 1, I to play the words $v_{i_1}, v_{i_2}, \ldots, v_{i_M}$. But since $u_{i_1} u_{i_2} \ldots u_{i_M} = v_{i_1} v_{i_2} \ldots v_{i_M}$, this is equivalent to playing the u_i words with the same indices, which is exactly the strategy that player 1, I uses.

To prove the converse, namely that if there is a winning strategy for coalition I then the correspondence problem has a solution, observe two intuitive facts. First, 2, I can never deduce in which component the play is taking place, because what he can see after each of his moves is the same in both components. Secondly, $\square$ can be played by 2, I only if the words played up to that point have the same length in both components. Otherwise, coalition I would lose as $\square$ can not be played in a subgame for any word.

Formally, let us first fix the only rational strategy for 2, II, namely that if a number L was the most recent action in the play, then 2, II plays v_L, and if there were other actions from $\{a, b\}^*$ taken after the last time a number L was played, then he plays u_L. Note that the above construction implies that player $2, \text{II}$ knows in which component the play takes place, even if the move of $3, \text{II}$ is not visible for him. With this strategy fixed, the condition that coalition I has a winning strategy for $\mathcal{G}_{\overline{u}, \overline{v}}$ means that there exists a strategy σ_1 for player $1, \text{I}$ and a strategy σ_2 for player $2, \text{I}$ such that the play corresponding to these two strategies and the one fixed for $2, \text{II}$ is winning for coalition I, independent of the component chosen by $3, \text{II}$.

Let us first concentrate on the strategy σ_2. Since, according to the winning condition, $\square$ can not be the only action played infinitely often, and in each component the only possible answer to $\square$ is again $\square$, let us assume without loss of generality that the first move taken by σ_2 is not $\square$ and let it be L_1. After choosing L_1 the play goes through v_{L_1} and u_{L_1} and does not stop, since player $2, \text{II}$ uses a fixed strategy that prevents him from losing in a subgame and player $1, \text{I}$ plays a winning strategy. Let us denote by L_2 the next move of $2, \text{I}$, i.e. $L_1 = \sigma_2(\varepsilon)$, $L_2 = \sigma_2(L_1 v_{L_1} u_{L_1})$, and continue the play denoting the subsequent moves of $2, \text{I}$ by $L_2, \ldots, L_M$, up to the point where he plays $\square$. Formally,

$$L_1 = \sigma_2(\varepsilon), \qquad L_{i+1} = \sigma_2(L_1 v_{L_1} u_{L_1} \ldots L_i v_{L_i} u_{L_i}), \qquad L_{M+1} = \square.$$

After extracting the sequence $L_1, \ldots, L_M$ of moves of $2, \text{I}$ from his winning strategy σ_2, let us look at player $1, \text{I}$. This is the only player on information level 1 so he only sees his own previous moves. In this case, the strategy σ_1 is in fact completely described by the word $t \in \{a, b, \square\}^\omega$ such that

$$t[i] = \sigma_1(t|_i) \text{ for all } i \in \mathbb{N}.$$

Due to the structure of the game, no $\square$ can be played by $1, \text{I}$ before $2, \text{I}$ decides to play $\square$, and then $\square$ must be played. Therefore, if w is the prefix of t up to the first occurrence of $\square$, then w is exactly the word played by $1, \text{I}$ while $2, \text{I}$ played the moves $L_1, \ldots, L_M$. But due to the structure of the game $\mathcal{G}_{\overline{u}, \overline{v}}$, coalition II can decide if $w = u_{L_1} \ldots u_{L_M}$ or if $w = v_{L_1} \ldots v_{L_M}$. Since we extracted both $L_1, \ldots, L_M$ and w independent of this choice, w has to be good for both cases. Therefore it is the solution for the Post correspondence problem as requested. $\square$

3.4 Model Checking with Hierarchical Games

We observed that non-alternating hierarchical games are neither determined nor decidable, so we concentrate on the alternating version. Indeed, we prove that alternating hierarchical games are exactly the games needed for model-checking FO[∂] on presentations of automatic structures.

To start with, observe that in an alternating game every infinite sequence of play actions can be divided into blocks of $2N$ actions, each taken by a different player,

$$\alpha = a_0^{1,I} a_0^{1,II} a_0^{2,I} a_0^{2,II} \ldots a_0^{N,I} a_0^{N,II} a_1^{1,I} \ldots a_1^{N,II} a_2^{1,I} \ldots .$$

Let the $2N$-*split* of these play actions be the tuple of $2N$ words of actions played by each of the players,

$$\text{split}_{2N}(\alpha) = (a_0^{1,I} a_1^{1,I} \ldots, \ \{a_i^{1,II}\}_{i \in \mathbb{N}}, \ \ldots, \ \{a_i^{N,I}\}_{i \in \mathbb{N}}, \ \{a_i^{N,II}\}_{i \in \mathbb{N}}).$$

Observe that since the set of plays winning for coalition I and starting from a fixed v_0 is ω-regular, also the set of corresponding $2N$-splits of play actions is ω-regular. This is a known property of ω-regular languages, and it can be proved by taking each $2N$th state of the automaton recognizing the plays and making a product with Σ^{2N} to store the states that were omitted from the original automaton. For an alternating hierarchical game $\mathcal{G}$ with winning condition $\mathcal{F}$ let us denote the $2N$ary relation recognizing the $2N$-split of plays winning for coalition I by $W_I^{\mathcal{G},v_0}(\beta_1, \ldots, \beta_{2N})$, formally defined by

$$W_I^{\mathcal{G},v_0}(\bar{\beta}) \iff \forall \alpha \ (\ \text{split}_{2N}(\alpha) = \bar{\beta} \ \Rightarrow \ \pi_\alpha(v_0) \in \mathcal{F} \ ).$$

The definition for coalition II is analogous with $\pi_\alpha(v_0)) \notin \mathcal{F}$.

Using the relation W_I^{G,v_0} we can express in $\mathsf{FO}[\partial]$ that coalition I wins in the alternating hierarchical game $\mathcal{G}$, which results in the following theorem.

Theorem 3.6. *For any alternating hierarchical game $\mathcal{G}$ and the relation $W_I^{\mathcal{G},v_0}$ defined as above, coalition I wins the game $\mathcal{G}$ starting from v_0 if and only if the following formula φ_I holds in $(\Sigma^\omega, W_I^{\mathcal{G},v_0})$:*

$$\varphi_I = \partial x_1 y_1 \ldots \partial x_N y_N \ W_I^{\mathcal{G},v_0}(x_1, y_1, \ldots, x_N, y_N).$$

Proof. Let us recapitulate the definition of coalition I winning a hierarchical game and the semantics of the formula φ_I. Coalition I wins $\mathcal{G}$ if there is a strategy σ_1 for player on level 1 of coalition I, so that for any counter-strategy ρ_1, there exists a strategy σ_2, and so on up to σ_N, such that for all ρ_N the resulting play must be won by coalition I. On the other hand, the formula φ_I, according to the definition of ∂, says that there is a function f_1, so that for all functions g_1, there is a function f_2, and so on up to a function f_N, such that for all g_N, if we construct the words according to $\bar{f}$ and $\bar{g}$ then they form a $2N$-split of a play that is won by coalition I.

As the structure and the final condition in both definitions are equivalent, due to the definition of $W_I^{\mathcal{G},v_0}$, the only remaining task is to show how the functions f_i, g_i and the strategies σ_i, ρ_i are related. It is intuitively clear that the functions and the strategies are closely related, the only difference is that the functions f_i, g_i operate on prefixes of x_i, y_i while the strategies σ_i, ρ_i take all actions of all players $j \leq i$ as arguments, which corresponds to prefixes of all words x_j, y_j with $j \leq i$. Intuitively, this makes no difference since the words x_j, y_j are completely fixed before the function f_i is constructed, and we are going to prove it formally.

Let us construct, given the function f_i, a strategy $\sigma_i^{f_i}$. The strategy $\sigma_i^{f_i}$ applied to a view h of the history of play actions extracts from h the sequences h_I^i and h_{II}^i of actions of players i, I and i, II, respectively, and chooses $f_i(h_I^i, h_{II}^i)$ as the next action. It is possible to extract h_I^i and h_{II}^i from h due to the alternation condition, because we know that h is of the form $a_0^{1,I} a_0^{1,II} a_0^{2,I} a_0^{2,II} \ldots a_0^{i,I} a_0^{i,II} a_1^{1,I} \ldots$ and the sequences $h_I^i = a_0^{i,I} a_1^{i,I} \ldots$ and h_{II}^i can be computed by taking every $2i$th position in h starting from $2i - 1$ and $2i$, respectively. Note that extracting these sequences would not be possible if it was not clear which player made which move, which we used in the previous proof of undecidability.

Let us now do the converse and construct, given the strategy σ_i, the function $f_i^{\sigma_i}$. For this construction we need to have all the f_j, g_j with $j < i$ already constructed, thus we write $f_i^{\{\sigma_j, \rho_j\}_{j \leq i}}$. Using the constructed functions f_j, g_j, we can assume that the words x_j, y_j are already fixed. The result of

$$f_i^{\{\sigma_j, \rho_j\}_{j \leq i}}(x_i[0] \ldots x_i[n], y_i[0], \ldots y_i[n])$$

is given by

$$\sigma_i(x_1[0]y_1[0] \ldots x_i[0]y_i[0]x_1[1]y_1[1] \ldots x_i[1]y_i[1] \ldots x_i[n]y_i[n]).$$

The constructions relating g_i and ρ_i are analogous. Observe that if

$$W_I^{\mathcal{G}, v_0}(x_{f_1 g_1} y_{f_1 g_1}, \ldots, x_{f_N g_N} y_{f_N g_N})$$

holds for some functions $\bar{f}, \bar{g}$ then, by the above definition, we have that the play $\pi(v_0, \sigma_1^{f_1}, \rho_1^{g_1}, \ldots, \sigma_N^{f_N}, \rho_N^{g_N})$ is in $\mathcal{F}$. Moreover, the converse holds as well, i.e. if for some strategies $\bar{\sigma}, \bar{\rho}$ we have

$$\pi(v_0, \sigma_1, \rho_1, \ldots, \sigma_N, \rho_N) \in \mathcal{F},$$

then $W_I^{\mathcal{G}, v_0}(x_{f_1 g_1} y_{f_1 g_1}, \ldots, x_{f_N g_N} y_{f_N g_N})$ holds, where $f_i = f_i^{\{\sigma_j, \rho_j\}_{j \leq i}}$ and $g_i = g_i^{\{\sigma_j, \rho_j\}_{j \leq i}}$ are the functions constructed above.

This correspondence allows to exploit the similarity of the structure of the definition of the FO[∂] formula φ_I and the definition of coalition I winning in $\mathcal{G}$. Intuitively, it is enough to insert the transformed functions and strategies into the definition to arrive at a contradiction and finish this proof. To avoid cluttered notation, we formally present only one direction in the case of two levels, the other direction and the proof for more levels is analogous.

Let us assume that φ_I holds and coalition I does not win $\mathcal{G}$, formally

(1) $\exists f_1 \forall g_1 \exists f_2 \forall g_2\ W_I^{\mathcal{G}, v_0}(x_{f_1 g_1}, y_{f_1 g_1}, x_{f_2 g_2}, y_{f_2 g_2}),$

(2) $\forall \sigma_1 \exists \rho_1 \forall \sigma_2 \exists \rho_2\ \pi(\sigma_1, \rho_1, \sigma_2, \rho_2) \notin \mathcal{F}.$

Let us fix f_1 that exists by our first assumption, set $\sigma_1 = \sigma_1^{f_1}$ and fix ρ_1 that exists by the second assumption for this σ_1. Let us now set $g_1 = g_1^{\rho_1}$ and fix

f_2 that exists by the first assumption. Finally, let us set $\sigma_2 = \sigma_2^{f_2}$ and fix ρ_2 that exists by the second assumption. By the previous observation

$$W_{\mathrm{I}}^{\mathcal{G},v_0}(x_{f_1g_1}, y_{f_1g_1}, x_{f_2g_2}, y_{f_2g_2}) \iff \pi(\sigma_1, \rho_1, \sigma_2, \rho_2) \in \mathcal{F},$$

but this contradicts the two assumptions above. □

Observe that the same proof works for the other coalition and an analogous relation $W_{\mathrm{II}}^{\mathcal{G},v_0}$. Thus, the negation normal form of $\mathsf{FO}[\partial]$ corresponds to determinacy of alternating hierarchical games.

Corollary 3.7. *Alternating hierarchical games are determined.*

After we captured winning in alternating games in $\mathsf{FO}[\partial]$ let us do the converse and construct the model-checking game for a given $\mathsf{FO}[\partial]$ formula on an automatic presentation $\mathfrak{A}$. At first, we restrict ourselves to formulas of the form

$$\varphi = \partial x_1 y_1 \partial x_2 y_2 \ldots \partial x_N y_N \; R(x_1, y_1, \ldots, x_N, y_N)$$

and construct a game so that the *split* of the winning plays will allow us to use the previous theorem.

Intuitively, the construction can be understood as prefixing each variable with all possible letters in the order of information hierarchy and making a step of the automaton when all the variables are prefixed. To define these games precisely, let us take the deterministic automaton for R, denoted $\mathcal{A}_R = (Q, q_0, \delta, \mathcal{F}_R)$, and construct the model-checking game $\mathcal{G}_\varphi$ for φ in the following way.

For each tuple of letters $c_1, d_1, c_2, d_2, \ldots, c_M, d_M$ of even length, with $0 \leq M < N$, and for every state $q \in Q$, we have in $\mathcal{G}_\varphi$ the position

$$R^q(c_1 x_1, d_1 y_1, \ldots, c_M x_M, d_M y_M, x_{M+1}, \ldots, y_N). \tag{3.1}$$

Moreover, for each tuple $c_1, d_1, c_2, d_2, \ldots, c_M, d_M, c_{M+1}$ of odd length, we have the position

$$R^q(c_1 x_1, \ldots, d_M y_M, c_{M+1} x_{M+1}, y_{M+1}, \ldots, y_N). \tag{3.2}$$

In each of these positions, the next move is made by the player corresponding to the next variable that is not yet prefixed by a letter, e.g. in position 3.1 it is the player $M + 1$ of coalition I who makes the move for x_{M+1} and in position 3.2 it is the player $M+1$ of coalition II. We can formally define the set of positions of players on each level i as $V_{i,\mathrm{I}} = Q \times \Sigma^{2(i-1)}$, $V_{i,\mathrm{II}} = Q \times \Sigma^{2i-1}$.

The moves in $\mathcal{G}_\varphi$ intuitively correspond to the player choosing a letter to prefix his variable with, so for $0 \leq M < N$

$$\mu(R^q(c_1 x_1, \ldots, d_M y_M, x_{M+1}, \ldots, y_N), c_{M+1}) =$$

$$R^q(c_1 x_1, \ldots, d_M y_M, c_{M+1} x_{M+1}, y_{M+1}, \ldots, y_N),$$

and for $0 \leq M < N - 1$

$$\mu(R^q(c_1x_1,\ldots,c_{M+1}x_{M+1},y_{M+1},\ldots,y_N),d_{M+1}) =$$

$$R^q(c_1x_1,\ldots,c_{M+1}x_{M+1},d_{M+1}y_{M+1},x_{M+2},\ldots,y_N).$$

The only special case is the final position $R^q(c_1x_1,d_1y_1,\ldots,c_Nx_N,y_N)$. When player N,II chooses the final letter d_N, it will not be appended, but instead all prefixing letters will be removed and the state of the automaton will be changed as follows, with $\overline{\alpha} = c_1d_1\ldots c_Nd_N$:

$$\mu(R^q(c_1x_1,d_1y_1,\ldots,c_Nx_N,y_N),d_N) = R^{\delta(q,\overline{\alpha})}(x_1,\ldots,y_N).$$

We derive the winning condition $\mathcal{F}$ of the game $\mathcal{G}_\varphi$ from the acceptance condition $\mathcal{F}_R$ of the automaton for R in the following way. Only the state component of each position in the game is taken into account, i.e. a sequence π of positions of $\mathcal{G}_\varphi$ is in $\mathcal{F}$ if and only if π projected to the state component is in $\mathcal{F}_R$.

To see that the game $\mathcal{G}_\varphi$ is indeed the model-checking game for φ, we use Theorem 3.6 and observe that the $2N$-split of the winning paths in $\mathcal{G}_\varphi$ is exactly the relation R, $W_{\mathrm{I}}^{\mathcal{G}_\varphi,R^{q_0}(x_1,y_1,\ldots,x_N,y_N)} = R$.

In this way, the model-checking game for formulas in the considered form is constructed. As we proved, any formula in $\mathsf{FO}[\eth]$ can be written in nega- tion normal form and additionally, by renaming variables, it can be trans- formed into prenex normal form. Let us therefore consider a general formula in the form $\varphi = \eth x_1y_1\ldots\eth x_Ny_N\ \psi(x_1,y_1,\ldots,x_N,y_N)$, where ψ is in nega- tion normal form and does not contain quantifiers. We construct the game $\mathcal{G}_\varphi$ inductively with respect to ψ.

In the case of $\psi(\overline{x}) = R(\overline{x})$ or $\psi(\overline{x}) = \neg R(\overline{x})$ the solution was already presented, when considering $\neg R$ we just have to complement the acceptance condition of the automaton for R. Let us show how to construct the game for Boolean connectives, i.e. for $\psi_1 \wedge \psi_2$ and for $\psi_1 \vee \psi_2$. We want to adhere to the usual convention of model-checking games and to have only one additional position for any Boolean connective. The game for $\psi_1 \circ \psi_2$, where $\circ = \wedge, \vee$, is therefore constructed as follows: we take the two games for ψ_1 and ψ_2 and we add one more position on higher level of information that has two possible moves — to the starting position of ψ_1 and to the starting position of ψ_2. The new position belongs to coalition I when $\circ = \vee$ and to coalition II when $\circ = \wedge$ and in both cases the other coalition does not play on that information level. With the construction described above we face a problem, as the game is not strictly alternating any more, but this time it can be made alternating by adding dummy positions, as presented in Example 3.8.

To formally prove that the resulting games are indeed model-checking games for formulas with Boolean connectives, we replace the connectives with a new variable and the formula with a relation where only the first letter of the new variable corresponding to the Boolean connective is considered. Then the automaton for such a relation corresponds to the defined game and Theorem 3.6 can be used again.

Example 3.8. To illustrate the construction of model-checking games and the method to overcome the technical problem with non-alternating games mentioned above, let us consider the simple formula $\exists x\ (R_1(x) \wedge R_2(x))$ over $\{a, b\}^\omega$ with $R_1 = \{a^\omega\}$ and $R_2 = \{a, b\}^\omega \setminus \{a^\omega\}$. Both the automaton for R_1 and the one for R_2 has two states and the transition functions are identical. On any b the automata go from q_0 to q_1 and stay there forever. Only the Büchi acceptance conditions differ, with $F_1 = \{q_0\}$ and $F_2 = \{q_1\}$.

In Figure 3.6, the game for this formula is depicted. We show dummy moves for the second player, as formally $\exists x\varphi(x) \equiv \partial xy\varphi(x)$. Note that this is actually a four-player game and the top position belongs to player 2, II. Since the formula is false, coalition II wins this game. Indeed, for coalition I to win, player 1, I would have to present a strategy to visit both of the double-circled vertices infinitely often without knowing in which branch he is, and that is impossible.

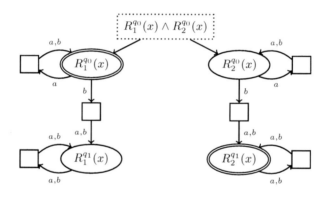

Fig. 3.6. Model-checking game for $\exists x(R_1(x) \wedge R_2(x))$

To fix the problem with alternation, let us add positions where there is no choice for the player. The alternating game for $\psi_1 \circ \psi_2$ is depicted in Figure 3.7. In this game, dummy positions are added there, where it is necessary to make the game alternating. It is clear that winning strategies in these two games can be transferred, as in each move on each level of visibility the players know how many moves on the other levels were made, both in the original game depicted in Figure 3.6 and in the modified one in Figure 3.7.

The tight correspondence between alternating hierarchical games and FO[∂] makes it possible to use our knowledge about this logic to reason about the games. In particular, we can transfer the results about complexity, including the non-elementary lower bound on deciding FO[∂] on automatic presentations, which allows us to conclude with the following corollary.

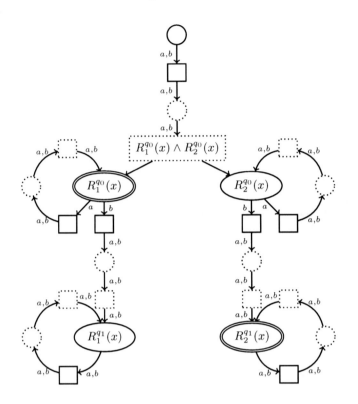

Fig. 3.7. Alternating game for $\exists x(R_1(x) \wedge R_2(x))$

Corollary 3.9. *The question whether coalition* I *wins in an alternating hierarchical game on a finite arena is decidable and has non-elementary complexity when the number of levels is not fixed. It can be decided in* 2kEXPTIME *for games with at most k levels.*

4

Memory Structures for Infinitary Games

In the previous chapters, we explored the connections between logic and games in a generic way, without relating to a specific representation of winning conditions in games. In this chapter, we investigate explicitly given winning conditions in terms of the complexity of strategies that are needed to win games with a fixed condition.

This question has been answered to a large extent for winning conditions defined over finite sets of priorities. We look at games with winning conditions defined over infinite sets of priorities and construct memory structures for different types of conditions in this case. Inspired by the notion of latest appearance record [39] used for games with finitely many priorities, we define the finite appearance record [34] and investigate which types of winning conditions are determined with such memory. The class of these conditions includes:

- downward cones,
- singleton conditions,
- finite unions of upwards cones,
- Muller conditions with finitely many winning sets,
- max-parity condition on graphs with bounded moves.

It remains open whether arbitrary max-parity games are determined via finite appearance records and a complete classification of Muller conditions over an infinite set of priorities with this property is not obtained. Still, the reduction for Muller conditions containing finitely many (possibly infinite) sets is a strong generalization of the classical case over finitely many priorities.

In addition to finite appearance records, we investigate winning conditions for which a certain representation, called the Zielonka tree [88], exists. These include all conditions over a finite set of priorities, for which the connection between the Zielonka tree and the memory needed for strategies is well understood [88]. We show that under certain assumptions this classical result can be transferred to infinite number of priorities as well.

4.1   Memory Structures and Determinacy

The most general representation of ω-regular acceptance conditions (for automata) and winning conditions (for games) that we considered so far was the Muller condition, which we defined as a class of subsets of states of the automaton or positions of the game. In this chapter, we study these conditions more thoroughly, and for this reason we give a slightly more general definition and extend the notation.

To start with, we assume that every game we consider has an arena labeled by priorities from a set C. Formally, a game is now not only the tuple $(V_0, V_1, E, \mathcal{F})$ but it consists of the *game graph* $G = (V, V_0, V_1, E)$ (with $V = V_0 \cup V_1$) which, together with a labeling function $\Omega : V \to C$, forms the *game arena* (G, Ω). A game is defined as a game arena together with a winning condition, $\mathcal{G} = (G, \Omega, \mathcal{F})$, and we focus on a Muller winning condition $\mathcal{F}$ over C defined as follows.

Definition 4.1. *A Muller condition over a finite set C of priorities is written in the form $(\mathcal{F}_0, \mathcal{F}_1)$ where $\mathcal{F}_0 \subseteq \mathcal{P}(C)$ and $\mathcal{F}_1 = \mathcal{P}(C) - \mathcal{F}_0$. A play π in a game with Muller winning condition $(\mathcal{F}_0, \mathcal{F}_1)$ is won by Player σ if, and only if, $\mathrm{Inf}(\pi)$, the set of priorities occurring infinitely often in π, belongs to $\mathcal{F}_\sigma$. A Streett-Rabin condition is now defined as a Muller condition $(\mathcal{F}_0, \mathcal{F}_1)$ such that $\mathcal{F}_0$ is closed under union.*

Definition 4.2. *A memory structure for a game $\mathcal{G}$ with positions in V is a triple $\mathfrak{M} = (M, \mathrm{update}, \mathrm{init})$, where M is a set of memory states, $\mathrm{update} : M \times V \to M$ is a memory update function and $\mathrm{init} : V \to M$ is a memory initialization function. The size of the memory is the cardinality of the set M. A strategy with memory $\mathfrak{M}$ for Player σ is given by a next-move function $F : V_\sigma \times M \to V$ such that $F(v, m) \in vE$ for all $v \in V_\sigma, m \in M$. If a play, from starting position v_0, has gone through positions $v_0 v_1 \ldots v_n$ the memory state is $m(v_0 \ldots v_n)$, defined inductively by $m(v_0) = \mathrm{init}(v_0)$, and $m(v_0 \ldots v_i v_{i+1}) = \mathrm{update}(m(v_0 \ldots v_i), v_{i+1})$. In case $v_n \in V_\sigma$, the next move from $v_1 \ldots v_n$, according to the strategy, leads to $F(v_n, m(v_0 \ldots, v_n))$. In case $|M| = 1$, the strategy is positional; it can be described by a function $F : V_\sigma \to V$.*

We will say that a game is determined via memory $\mathfrak{M}$ if it is determined and both players have winning strategies with memory $\mathfrak{M}$ on their winning regions. A game is positionally determined if it is determined via positional winning strategies.

Given a game graph $G = (V, V_0, V_1, E)$ and a memory structure $\mathfrak{M} = (M, \mathrm{update}, \mathrm{init})$ we obtain a new game graph $G \times \mathfrak{M} = (V \times M, V_0 \times M, V_1 \times M, E_{\mathrm{update}})$ where

$$E_{\mathrm{update}} = \{(v, m)(v', m') : (v, v') \in E \text{ and } m' = \mathrm{update}(m, v')\}.$$

Obviously, every play $(v_0, m_0)(v_1, m_1) \ldots$ in $G \times \mathfrak{M}$ has a unique projection to the play $v_0 v_1 \ldots$ in G. Conversely, every play $v_0, v_1, \ldots$ in G has a unique extension to a play $(v_0, m_0)(v_1, m_1) \ldots$ in $G \times \mathfrak{M}$ with $m_0 = \text{init}(v_0)$ and $m_{i+1} = \text{update}(m_i, v_{i+1})$.

Consider two games $\mathcal{G} = (G, \Omega, W)$ and $\mathcal{G}' = (G', \Omega', W')$. We say that $\mathcal{G}$ *reduces via memory* $\mathfrak{M}$ *to* $\mathcal{G}'$, (in short $\mathcal{G} \leq_{\mathfrak{M}} \mathcal{G}'$) if $G' = G \times \mathfrak{M}$ and every play in $\mathcal{G}'$ is won by the same player as the projected play in $\mathcal{G}$.

Given a memory structure $\mathfrak{M}$ for G and a memory structure $\mathfrak{M}'$ for $G \times \mathfrak{M}$ we obtain a memory structure $\mathfrak{M}^* = \mathfrak{M} \times \mathfrak{M}'$ for G. The set of memory locations is $M \times M'$ and we have memory initialization $\text{init}^*(v) = (\text{init}(v), \text{init}'(v, \text{init}(v)))$ and the update function

$$\text{update}^*((m, m'), v) =$$
$$= (\text{update}(m, v), \text{update}'(m', (v, \text{update}(m, v)))).$$

Proposition 4.3. *Suppose that a game $\mathcal{G}$ reduces to $\mathcal{G}'$ via memory $\mathfrak{M}$ and that Player σ has a winning strategy for $\mathcal{G}'$ with memory $\mathfrak{M}'$ from $(v_0, \text{init}(v_0)))$. Then Player σ has a winning strategy for $\mathcal{G}$ with memory $\mathfrak{M} \times \mathfrak{M}'$ from position v_0.*

Proof. Given a strategy $F' : (V_\sigma \times M) \times M' \to (V \times M)$ for Player σ on $\mathcal{G}'$ we have to construct a strategy $F : (V_\sigma \times (M \times M')) \to V \times (M \times M')$.

For a pair $(v, m) \in V_\sigma \times M$, we have that $F'(v, m) = (w, \text{update}(m, w))$ where $w \in vE$. We now put $F(v, mm') = w$. If a play in $\mathcal{G}$ that is consistent with F proceeds from position v, with current memory location (m, m'), to a new position w, then the memory is updated to (n, n') with $n = \text{update}(m, w)$ and $n' = \text{update}'(m', (w, n))$. In the extended play in $\mathcal{G}'$ we have an associated move from position (v, m) to (w, n) with memory update from m' to n'. Thus, every play in $\mathcal{G}$ from initial position v_0 that is consistent with F is the projection of a play in $\mathcal{G}'$ from $(v_0, \text{init}(v_0))$ that is consistent with F'. Therefore, if F' is a winning strategy from $(v_0, \text{init}(v_0))$, then F is a winning strategy from v_0. $\square$

Corollary 4.4. *Every game that reduces via memory $\mathfrak{M}$ to a positionally determined game, is determined via memory $\mathfrak{M}$.*

Obviously, memory reductions between games compose. If $\mathcal{G}$ reduces to $\mathcal{G}'$ with memory $\mathfrak{M}$ and $\mathcal{G}'$ reduces to $\mathcal{G}''$ with memory $\mathfrak{M}'$ then $\mathcal{G}$ reduces to $\mathcal{G}''$ with the memory $\mathfrak{M}^* = \mathfrak{M} \times \mathfrak{M}'$ defined above.

4.2 Latest Appearance Record for Muller Games

One of the reasons for the interest in parity games is the fact that parity games over a finite set of priorities $C = \{0, \ldots, d\}$ are positionally determined [31, 68]. The classical example of a game reduction with finite memory on the other hand is the reduction of Muller games to parity games via latest appearance records [39]. Intuitively, a latest appearance record (LAR) is a list of priorities ordered by their latest occurrence.

More formally, for a finite set C of priorities, $\mathsf{LAR}(C)$ is the set of sequences $c_1 \ldots c_k \natural c_{k+1} \ldots c_\ell$ of elements from $C \cup \{\natural\}$ in which each priority $c \in C$ occurs at most once, and $\natural$ occurs precisely once. At a position v, the LAR $c_1 \ldots c_k \natural c_{k+1} \ldots c_\ell$ is updated by moving the priority $\Omega(v)$ to the end, and moving $\natural$ to the previous position of $\Omega(v)$ in the sequence. For instance, at a position with priority c_2, the LAR $c_1 c_2 c_3 \natural c_4 c_5$ is updated to $c_1 \natural c_3 c_4 c_5 c_2$. (If $\Omega(v)$ did not occur in the LAR, we simply append $\Omega(v)$ at the end). Thus, the LAR-memory for an arena with priority labeling $\Omega : V \to C$ is the triple $(LAR(C), \mathrm{init}, \mathrm{update})$ with $\mathrm{init}(v) = \natural \Omega(v)$ and

$$\mathrm{update}(c_1 \ldots c_k \natural c_{k+1} \ldots c_\ell, v) = c_1 \ldots c_k \natural c_{k+1} \ldots c_\ell \Omega(v)$$

in case $\Omega(v) \notin \{c_1 \ldots c_\ell\}$, and

$$\mathrm{update}(c_1 \ldots c_k \natural c_{k+1} \ldots c_\ell, v) = c_1 \ldots c_{m-1} \natural c_{m+1} \ldots c_\ell c_m$$

if $\Omega(v) = c_m$.

The *hit-set* of a LAR $c_1 \ldots c_k \natural c_{k+1} \ldots c_\ell$ is the set $\{c_{k+1} \ldots c_\ell\}$ of priorities occurring after the symbol $\natural$. Observe that if in a play $\pi = v_0 v_1 \ldots$, the LAR at position v_n is $c_1 \ldots c_k \natural c_{k+1} \ldots c_\ell$ then $\Omega(v_n) = c_\ell$ and the hit-set $\{c_{k+1} \ldots c_\ell\}$ is the set of priorities that have been seen since the latest previous occurrence of c_ℓ in the play.

Lemma 4.5. *Let π be a play of a Muller game $\mathcal{G}$, and let $\mathrm{Inf}(\pi)$ be the set of priorities occurring infinitely often in π. On π the hit-set of the latest appearance record is, from some point onwards, always a subset of $\mathrm{Inf}(\pi)$ and infinitely often coincides with $\mathrm{Inf}(\pi)$.*

Proof. For each play $\pi = v_0 v_1 v_2 \ldots$ there is a position v_m such that $\Omega(v_n) \in \mathrm{Inf}(\pi)$ for all $n \geq m$. Since no priority outside $\mathrm{Inf}(\pi)$ is seen anymore after position v_m, the hit-set will from that point onwards always be contained in $\mathrm{Inf}(\pi)$, and the LAR will always have the form $c_1 \ldots c_{j-1} c_j \ldots c_k \natural c_{k+1} \ldots c_\ell$ where $c_1, \ldots c_{j-1}$ remain fixed and the set $\{c_j, \ldots, c_k, c_{k+1}, \ldots c_\ell\} = \mathrm{Inf}(\pi)$. Since all priorities in $\mathrm{Inf}(\pi)$ are seen again and again, it happens infinitely often that, among these, the one occurring leftmost in the LAR is hit. At such positions, the LAR is updated to $c_1, \ldots, c_{j-1} \natural c_{j+1} \ldots c_\ell c_j$ and the hit-set then coincides with $\mathrm{Inf}(\pi)$. □

Theorem 4.6. *Every Muller game with finitely many priorities reduces via LAR memory to a parity game.*

Proof. Let $\mathcal{G}$ be a Muller game with game graph G, priority labeling $\Omega : V \to C$ and winning condition $(\mathcal{F}_0, \mathcal{F}_1)$. We have to prove that $\mathcal{G} \leq_{\mathsf{LAR}} \mathcal{G}'$ for a parity game $\mathcal{G}'$ with game graph $G \times \mathsf{LAR}(C)$ and an appropriate priority labeling Ω' on $V \times \mathsf{LAR}(C)$ which is defined as follows.

$$\Omega'(v, c_1 c_2 \ldots c_k \natural c_{k+1} \ldots c_\ell) = \begin{cases} 2k & \text{if } \{c_{k+1}, \ldots, c_\ell\} \in \mathcal{F}_0, \\ 2k+1 & \text{if } \{c_{k+1}, \ldots, c_\ell\} \in \mathcal{F}_1. \end{cases}$$

Let $\pi = v_0 v_1 v_2 \ldots$ be a play on $\mathcal{G}$ and fix a number m such that, for all numbers $n \geq m$ and $\Omega(v_n) \in \mathrm{Inf}(\pi)$, the LAR at position v_n has the form $c_1 \ldots c_j c_{j+1} \ldots c_k \natural c_{k+1} \ldots c_\ell$ where $\mathrm{Inf}(\pi) = \{c_{j+1}, \ldots c_\ell\}$ and the prefix $c_1 \ldots c_j$ remains fixed. In the extended play $\pi' = (v_0 r_0)(v_1, r_1) \ldots$ all nodes (v_n, r_n) for $n \geq$ will therefore have a priority $2k + \rho$ with $k \geq j$ and $\rho \in \{0, 1\}$. Assume that the play π is won by Player σ, i.e. $\mathrm{Inf}(\pi) \in \mathcal{F}_\sigma$. Since infinitely often the hit-set of the LAR coincides with $\mathrm{Inf}(\pi)$, the minimal priority seen infinitely often on the extended play is $2j + \sigma$. Thus the extended play in the parity game $\mathcal{G}'$ is won by the same player as the original play in the Muller game $\mathcal{G}$.   □

Observe that for a Muller game on n priorities, an LAR-memory has $n!$ memory states. Dziembowski, Jurdziński, and Walukiewicz [26] have shown that with this respect LAR-strategies are essentially optimal for Muller games.

Theorem 4.7. *There exists a sequence $(\mathcal{G}_n)_{n \in \omega}$ of Muller games such that the game graph of $\mathcal{G}_n$ is of size $O(n)$ and every winning strategy for $\mathcal{G}_n$ requires a memory of size at least $n!$*

4.3 Games with Infinitely Many Priorities

The definition of the Muller condition (Definition 4.1) directly generalizes to countable sets C of priorities. Note that with minor modifications it can also be generalized to uncountable sets C, see [35] for a discussion of this. But some properties that hold for a finite set of priorities C do not generalize even to countable sets. One of them is the possibility to represent any Muller condition by a Zielonka tree, which we discuss in section 4.6.

For finitely many priorities, the condition that $\mathcal{F}_0$ and $\mathcal{F}_1$ are both closed under finite unions is sufficient for positional determinacy of any game with this Muller condition. To see that this is not the case for infinite sets C, let us discuss the possible generalizations of parity games to the case of priority assignments $\Omega : V \to \omega$. For parity games with finitely many priorities it is of course purely a matter of taste whether we let the winner be determined by the least priority seen infinitely often or by the greatest one. Here this is no longer the case. Based on priority assignments $\Omega : V \to \omega$, we consider the following classes of games.

Infinity games are games where Player 0 wins those infinite plays in which no priority at all appears infinitely often, i.e.

$$\mathcal{F}_0 = \{\emptyset\},$$
$$\mathcal{F}_1 = \mathcal{P}(\omega) \setminus \{\emptyset\}.$$

Parity games are games where Player 0 wins the plays in which the least priority seen infinitely often is even, or where no priority appears infinitely often. Thus,

$$\mathcal{F}_0 = \{X \subseteq \omega : \min(X) \text{ is even}\} \cup \{\emptyset\},$$
$$\mathcal{F}_1 = \{X \subseteq \omega : \min(X) \text{ is odd}\}.$$

Max-parity games are games where Player 0 wins if the maximal priority occurring infinitely often is even, or does not exist, i.e.

$$\mathcal{F}_0 = \{X \subseteq \omega : \text{ if } X \neq \emptyset \text{ and } X \text{ is finite then } \max(X) \text{ is even}\},$$
$$\mathcal{F}_1 = \{X \subseteq \omega : X \text{ is finite, non-empty, and } \max(X) \text{ is odd}\}.$$

It is easy to see that infinity games are a special case of parity games, via a simple reassignment of priorities. Further, we note that for both parity games and max-parity games, $\mathcal{F}_0$ and $\mathcal{F}_1$ are closed under finite unions. Nevertheless the conditions behave quite differently.

Proposition 4.8. *Max-parity games with infinitely many priorities in general do not admit finite memory winning strategies.*

Proof. Consider the max-parity game with positions $V_0 = \{0\}$ and $V_1 = \{2n + 1 : n \in \mathbb{N}\}$ (where the name of a position is also its priority), such that Player 0 can move from 0 to any position $2n + 1$ and Player 1 can move back from $2n+1$ to 0. Clearly Player 0 has a winning strategy from each position but no winning strategy with finite memory. □

On the other hand it has been shown in [35] that infinity games and parity games with priorities in ω do admit positional winning strategies for both players on all game graphs. In fact, parity games over ω turn out to be the only Muller games with this property.

Theorem 4.9. *[35] Let $(\mathcal{F}_0, \mathcal{F}_1)$ be a Muller winning condition over a countable set C of priorities. Then the following are equivalent.*

- *Every game with winning condition $(\mathcal{F}_0, \mathcal{F}_1)$ is positionally determined.*
- *Both $\mathcal{F}_0$ and $\mathcal{F}_1$ are closed under finite unions, unions of chains, and non-empty intersections of chains.*
- *The Zielonka tree of $(\mathcal{F}_0, \mathcal{F}_1)$ exists, and is a path of co-finite sets (and possibly the empty set at the end).*
- *$(\mathcal{F}_0, \mathcal{F}_1)$ reduces to a parity condition over $n \leq \omega$ priorities.*

4.4 Finite Appearance Records

Although over an infinite set of priorities one can easily define Muller games that do not admit finite memory strategies, these games are often solvable by strategies with very simple infinite memory structures. For instance, for the max-parity game described in the proof of Proposition 4.8, it suffices for Player 0 to store the maximal priority seen so far, in order to determine the next move in her winning strategy. One can readily come up with other games where the memory required by a winning strategies is essentially a finite collection of previously seen priorities.

This motivates the definition of an infinite memory structure that we call *finite appearance records* (FAR) which generalizes the LAR-memory for games with finitely many priorities. In a FAR we store tuples of previously encountered priorities or some other symbols from a finite set. Additionally the update function in the appearance record is restricted, so that new values of the memory can be equal only to the values stored before or to the currently seen priority.

Definition 4.10. *A d-dimensional FAR-memory for a game $\mathcal{G}$ with priorities in C is a memory structure $(M, \text{update}, \text{init})$ for $\mathcal{G}$ with $M = (C \cup N)^d$ for some finite set N such that whenever*

$$\text{update}(m_1, \ldots, m_d, v) = (m'_1, \ldots, m'_d)$$

then $m'_i \in \{m_1, \ldots, m_d\} \cup N \cup \{\Omega(v)\}$.

Observe that an LAR-memory over a finite set C is a special case of an FAR-memory, with $d = |C| + 1$ and $N = \{\natural, B\}$, where B is a blank symbol used to pad latest appearance records in which some priorities are missing. Here the dimension of the FAR depends on the size of C. Hence, the question arises whether there is a fixed dimension d and a fixed additional set N such that every Muller game over finitely many priorities reduces to a parity game via d-dimensional FAR-memory. From Theorem 4.7 it follows that his is not the case. Indeed, since $n!$ grows faster than n^d for any constant d, we infer that for any dimension d there is a Muller game $\mathcal{G}_d$ that can not be reduced to a parity game via d-dimensional FAR-memory. From this we obtain the following conclusion.

Proposition 4.11. *There exists a Muller game $\mathcal{G}$ that does not reduce to a parity game with any FAR-memory.*

Proof. Take $\mathcal{G}$ to be the disjoint sum of the games $\mathcal{G}_d$, assuming that all these games have disjoint sets of priorities. Suppose that $\mathcal{G}$ reduces to a parity game via some FAR-memory of dimension d. Since game extensions preserve connectivity it follows that the extension of the connected component $\mathcal{G}_d$ of $\mathcal{G}$ will also be a parity game. But this contradicts the fact that $\mathcal{G}_d$ does not reduce to a parity game via d-dimensional FAR-memory.   □

4.5 FAR Reductions for Infinitary Muller Games

In this section we consider some cases of Muller games with priorities in ω that admit FAR-reductions to positionally determined games.

To illustrate the idea consider any *downwards cone* $\mathcal{F}_0 = \{X : X \subseteq A\}$ for a fixed set $A \subseteq \omega$. Again it is easy to see that such games may require infinite-memory strategies. To reduce such a game to a parity game $\mathcal{G}'$ it suffices to store the maximal priority m seen so far, and to define priorities in $\mathcal{G}'$ by

$$\Omega'(v,m) = \begin{cases} 2m + 2 & \text{if } \Omega(v) \in A, \\ 2\Omega(v) + 1 & \text{otherwise.} \end{cases}$$

If $\text{Inf}(\pi) \subseteq A$ then Player 0 wins π' since no odd priority is seen infinitely often in π'. If there is some $a \in \text{Inf}(\pi) \setminus A$, then $2a + 1$ occurs infinitely often in π', and since $a \leq m$ from some point onwards, no smaller even priority can have this property, so Player 1 wins π'.

Hence any Muller game such that $\mathcal{F}_0$ (or $\mathcal{F}_1$) is a downwards cone is determined via one-dimensional FAR-memory.

4.5.1 Visiting Sequences and Singleton Muller Conditions

Our next example for winning conditions that are amenable for an approach via FAR-reductions are Muller games where the winning condition of Player 0 is a singleton, i.e. $\mathcal{F}_0 = \{A\}$, $\mathcal{F}_1 = \mathcal{P}(\omega) \setminus \{A\}$.

We first observe that such games may require infinite memory.

Theorem 4.12. *For any $A \neq \emptyset$, there exists a (solitaire) Muller game with $\mathcal{F}_0 = \{A\}$ whose winning strategies all require infinite memory.*

Proof. If $A = \{a_1, a_2, \dots\}$ is infinite, take the game with set of positions $V = V_0 = A$ (where the name of a position indicates also its priority), and moves (a_1, a_n) and (a_n, a_1) for all $n \geq 2$. If $A = \{a_1, \dots, a_n\}$ is finite, let $\omega \setminus A = \{b_1, b_2, \dots\}$. We consider instead the game with $V = V_0 = A \cup (\omega \setminus A)$, and set of moves

$$E = \{(a_i, a_{i+1}) : 1 \leq i < n\} \cup$$
$$\{(a_n, b) : b \in (\omega \setminus A)\} \cup \{(b, a_1) : b \in (\omega \setminus A)\}.$$

In both cases, Player 0 wins, but requires infinite memory to do so.   □

We will prove that singleton Muller games can be reduced via FAR-memory to parity games with priorities in ω which, as shown in [35], are positionally determined. The FAR-memory that we use for this reduction is based on a particular order in which the elements of the winning sets have to be seen infinitely often, which is specified by a visiting sequence.

Definition 4.13. *Let $A = \{a_1 < a_2 < \dots\}$ be an infinite subset of ω. For each $n \in \omega$, let $p(a_n) := a_1 a_2 \dots a_n$ be the prefix of a_n. The visiting sequence of A is the concatenation of the prefixes of all elements of A*

$$\text{visit}(A) = p(a_1)p(a_2)p(a_3)\dots$$

For a finite set $\{a_1 < a_2 < \dots < a_n\} \subseteq \omega$ we define $\text{visit}(A) = p(a_n)^\omega$.

Let $\mathcal{G}$ be a Muller game over ω.

Lemma 4.14. *For any play* $\pi = v_1 v_2 \ldots$ *of* $\mathcal{G}$ *the set* $\mathrm{Inf}(\pi)$ *is the unique set* A *with the following two properties:*

(1) There exists a sequence of indices $i_1 < i_2 < \ldots$ *such that the priorities* $\Omega(v_{i_1})\Omega(v_{i_2})\ldots$ *form the visiting sequence of* A.
(2) If $\Omega(v_k) \in \omega \setminus A$ *then there is only a finite number of indices* $i > k$ *such that* $\Omega(v_i) \in \{0, \ldots, \Omega(v_k)\} \cap \omega \setminus A$.

Proof. First we notice that $A = \mathrm{Inf}(\pi)$ indeed fulfills these two properties. The visiting sequence can be chosen from the play as all elements of $\mathrm{Inf}(\pi)$ appear infinitely often. Since all elements of $\omega \setminus \mathrm{Inf}(\pi)$ occur only finitely often in the play, the second property must also hold.

Conversely, if a set A satisfies property (1), then all elements of A appear infinitely often in π, so $A \subseteq \mathrm{Inf}(\pi)$. If there were an element $a \in \mathrm{Inf}(\pi) \setminus A$, then for any k with $\Omega(v_k) = a$, there were infinitely many indices $i > k$, with $\Omega(v_i) = a$ which contradicts property (2). Thus if A satisfies properties (1) and (2), then $A = \mathrm{Inf}(\pi)$. □

Let $A \subseteq \omega$ be infinite. Any initial segment of the visiting sequence of A can be written in the form $p(a_1)p(a_2)\ldots p(a_i)a_1 a_2 \ldots a_j$ where $1 \leq j \leq i+1$. It can be represented by a pair (p,c) where $c = a_j$ indicates the position of the last letter in the current prefix $p(a_{i+1})$, and $p = a_i$ indicates the last previously completed prefix (or ε if we are at the first element). For instance, the initial segment $a_1 \, a_1 a_2 \, a_1 a_2 a_3 \, a_1 a_2 a_3$ of the visiting sequence of A is encoded by (a_3, a_3), the initial segment a_1 is encoded by (ε, a_1), and the empty initial segment by $(\varepsilon, \varepsilon)$. We write $\mathrm{visit}_n(A)$ for the initial segment of length A of $\mathrm{visit}(A)$.

Given a (finite or infinite) winning set A, we want to use a three-dimensional FAR-memory to check whether $\mathrm{Inf}(\pi) = A$. For infinite A, the memory state after an initial segment of a play is a triple (p, c, q) where (p, c) encode the initial segment of the visiting sequence of A that has been seen so far, and q is the maximal priority that has occurred.

Definition 4.15. *For any infinite set* $A \subseteq \omega$, *we define a three-dimensional FAR-memory* $\mathrm{FAR}(A) = (M, \mathrm{init}, \mathrm{update})$ *with* $M = \{(p, c, q) : p, c \in \omega \cup \{\varepsilon\}, q \in \omega\}$. *The initialization function is defined by*

$$\mathrm{init}(v) = \begin{cases} (\varepsilon, \Omega(v), \Omega(v)) & \text{if } \Omega(v) = a_1 \\ (\varepsilon, \varepsilon, \Omega(v)) & \text{if } \Omega(v) \neq a_1 \end{cases}$$

The update function is defined by

$$\mathrm{update}(p, c, q, v) = (p', c', q'),$$

where $q' = \max(q, \Omega(v))$, *and where* (p, c) *and* (p', c') *encode, for some* n, *the initial segments* $\mathrm{visit}_n(A)$ *and* $\mathrm{visit}_{n+1}(A)$, *respectively, of the visiting sequence of* A *such that* $\mathrm{visit}_{n+1}(A) = \mathrm{visit}_n(A)\Omega(v)$, *or otherwise,* $(p', c') = (p, c)$.

For a more formal description, let

$$\text{up}(p, c, v) = \begin{cases} 2 & \text{if, for some } i, \, p = a_i, c = a_{i+1}, \Omega(v) = a_1 \\ 1 & \text{if, for some } j \leq i, \, p = a_i, c = a_j, \Omega(v) = a_{j+1} \\ 0 & \text{otherwise} \end{cases}$$

(where, to simplify notation, we identify ε with a_0). Note that $\text{up}(p, c, v) = 2$ if, at node v, the visiting sequence is updated with an a_1 (i.e. a prefix $p(a_i)$ has been completed and a new one is started), that $\text{up}(p, c, v) = 1$ if the visiting sequence is updated by another value, and that $\text{up}(p, c, v) = 0$ if no update of the visiting sequence happens at v. Then we can define $\text{update}(p, c, q, v) := (p', c', q')$ by

$$(p', c') = \begin{cases} (c, \Omega(v)) & \text{if } \text{up}(p, c, v) = 2 \\ (p, \Omega(v)) & \text{if } \text{up}(p, c, v) = 1 \\ (p, c) & \text{if } \text{up}(p, c, v) = 0 \end{cases}$$

$$q' = \max(q, \Omega(v))$$

For finite $A = \{a_1 < a_2 < \cdots < a_n\}$ this has to be modified since once cannot really encode the part of the visiting sequence that one has seen with priorities in A. In this case the value (p, c, q) is so that c is the last element of the visiting sequence, q is the maximal priority that has occurred so far, and p is the maximal priority that had occurred up to the last time when, in the visiting sequence of A, a prefix $p(a_n)$ had been completed and c had been updated from a_n to a_1. Thus we set

$$\text{up}(p, c, v) = \begin{cases} 2 & \text{if } c = a_n, \Omega(v) = a_1 \\ 1 & \text{if, for some } i < n, c = a_i, \Omega(v) = a_{i+1} \\ 0 & \text{otherwise} \end{cases}$$

and $\text{update}(p, c, q, v) := (p', c', q')$ with

$$(p', c') = \begin{cases} (q, \Omega(v)) & \text{if } \text{up}(p, c, v) = 2 \\ (p, \Omega(v)) & \text{if } \text{up}(p, c, v) = 1 \\ (p, c) & \text{if } \text{up}(p, c, v) = 0 \end{cases}$$

$$q' = \max(q, \Omega(v)).$$

Theorem 4.16. *Any singleton Muller game with $\mathcal{F}_0 = \{A\}$ can be reduced, via memory* $\text{FAR}(A)$, *to a parity game.*

Proof. The given Muller game $\mathcal{G}$ with arena (G, Ω) and Muller condition such that $\mathcal{F}_0 = \{A\}$ is reduced via memory $\text{FAR}(A)$ to a parity game $\mathcal{G}'$ with priority function $\Omega' : V \times \text{FAR}(A) \to \omega$ defined as follows.

$$\Omega'(v, p, c, q) = \begin{cases} 2p + 2 & \text{if } \Omega(v) \in A, \text{up}(p, c, v) \in \{1, 2\} \\ 2p + 3 & \text{if } \Omega(v) \in A, \text{up}(p, c, v) = 0 \\ \min(2p + 3, 2\Omega(v) + 1) & \text{if } \Omega(v) \notin A \end{cases}$$

We have to prove that any play $\pi = v_0 v_1 v_2 \ldots$ of $\mathcal{G}$ is won by the same player as the extended play $\pi' = (v_0, p_0, c_0, q_0)(v_1, p_1, c_1, q_1) \ldots$ of $\mathcal{G}'$.

We first assume that $\text{Inf}(\pi) = A$ and prove that either no priority at all occurs infinitely often in π' or the minimal such is even. If A is infinite, then the sequence of the values p_n diverges and therefore no priority will be seen infinitely often in π'. If A is finite then it may be the case that the sequence $(p_n)_{n \in \omega}$ converges, i.e. $p_n = p$ from some point onwards. But since the visiting sequence will be updated again and again this means that infinitely often the priority $2p+2$ occurs in π', and the only other priority that may occur infinitely often is $2p + 3$. Hence Player 0 wins π'.

For the converse, we assume that Player 1 wins π. We distinguish several cases. If there exist some $a \in A \setminus \text{Inf}(\pi)$ then from some point onwards, the visiting sequence cannot be updated anymore, so the sequence $(p_n)_{n \in \omega}$ stabilizes at some value p. Then the minimal priority seen infinitely often is either $2p + 3$, or $2\Omega(v) + 1$ for some $\Omega(v) \in \omega \setminus A$ and Player 1 also wins π'. If no such element a exists, then $A \subsetneq \text{Inf}(\pi)$ and there is a minimal element $b \in \text{Inf}(\pi) \setminus A$. If the sequence $(p_n)_{n \in \omega}$ diverges (which is always the case for infinite winning sets A) then the minimal priority seen infinitely often in π' is $2b + 1$. If A is finite then the sequence p_n may stabilize at some value p which coincides with the largest priority ever occurring in π. Hence $b \leq p$ and therefore $2b + 1 < 2p + 2$, so the minimal priority seen infinitely often in π' is $2b + 1$. Again Player 1 wins the associated play in the parity game. $\qquad\square$

Corollary 4.17. *Singleton Muller games are determined with* FAR *memory.*

4.5.2 Finite Unions of Upwards Cones

Visiting sequences can also be used for the case where $\mathcal{F}_0$ is a finite union of upwards cones, i.e.

$$\mathcal{F}_0 = \bigcup_{i=1}^{k} \{X : A_i \subseteq X \subseteq \omega\}$$

for some finite collection of sets $A_1, \ldots, A_k$.

The FAR-memory stores the pairs (p_i, c_i) encoding the visiting sequences of $A_1, \ldots, A_k$. All that has to checked is whether $A_i \subseteq \text{Inf}(\pi)$ for some i, which is the case if, and only if, one of the visiting sequences is updated infinitely often. Thus we can define priorities by

$$\Omega'(v, p_1, c_1, \ldots, p_k, c_k) = \begin{cases} 0 & \text{if } \text{up}(p_i, c_i, v) = 2 \text{ for some } i \\ 1 & \text{otherwise.} \end{cases}$$

Theorem 4.18. *Any Muller game such that $\mathcal{F}_\sigma$ is a finite union of upwards cones is determined via* FAR*-memory.*

4.5.3   Muller Conditions with Finitely Many Winning Sets

We now consider the case of Muller games whose winning conditions are defined by a finite collection of (possibly infinite) sets, $\mathcal{F}_0 = \{A_1, \ldots, A_k\}$. To extend the idea presented above to this case we are going to use the memory $\mathsf{FAR}(A_i)$ for each set A_i and additionally we have to remember when the set A_i is *active*, as is described below. The property of being active is stored in a value $a_i \in \{0, 1, 2\}$.

Definition 4.19. *For a finite collection* $\{A_1, \ldots, A_k\}$ *of sets* $A_i \subseteq \omega$, *we define a 4k-dimensional* FAR*-memory* $\mathsf{FAR}(A_1, \ldots, A_k) = (M, \mathrm{init}, \mathrm{update})$. *We denote the* FAR*-memory of* A_i *by* $\mathsf{FAR}(A_i) = (M_i, \mathrm{init}_i, \mathrm{update}_i)$. *Then* $M = M_1 \times M_2 \times \ldots \times M_k \times \{0, 1, 2\}^k$. *The initialization function is defined by*

$$\mathrm{init}(v) = (\mathrm{init}_1(v), \ldots, \mathrm{init}_k(v), \bar{0}).$$

The update function is defined by

$$\mathrm{update}(m_1, \ldots, m_k, a_1, \ldots, a_k, v) = $$
$$(\mathrm{update}_1(m_1, v), \ldots, \mathrm{update}_k(m_k, v), a_1', \ldots, a_k'),$$

where a_i' *is the new activation value for sequence* i *defined by*

$$a_i' = \begin{cases} 0 & \text{if } v \notin A_i \text{ and for some } j \leq k \; \mathrm{up}_j(m_j, v) > 0 \\ \min(2, a_i + 1) & \text{if } \mathrm{up}_i(m_i, v) = 2 \\ a_i & \text{otherwise.} \end{cases}$$

Theorem 4.20. *Any Muller game with* $\mathcal{F}_0 = \{A_1, \ldots, A_k\}$ *can be reduced, via memory* $\mathsf{FAR}(A_1, \ldots, A_k)$, *to a parity game.*

Proof. The given Muller game $\mathcal{G}$ with arena (G, Ω) and Muller condition such that $\mathcal{F}_0 = \{A_1, \ldots, A_k\}$ is reduced to a parity game $\mathcal{G}'$ with priority function Ω' defined by

$$\Omega'(v, \overline{m}, \overline{a}) = \begin{cases} \max_{\mathrm{Act}(v, \overline{a})}(2kp_i + 2r_i + 2) & \text{if exists } j \text{ such that} \\ & \Omega(v) \in A_j, a_j = 2, \\ & \mathrm{up}_j(m_j, v) \in \{1, 2\} \\ \max_{\mathrm{Act}(v, \overline{a})}(2kp_i + 2r_i + 3) & \text{if exists } j \text{ such that} \\ & \Omega(v) \in A_j, a_j = 2, \\ & \text{and } \mathrm{up}_j(m_j, v) = 0 \\ & \text{for all such } j \\ \min(2kp_{\max} + 3, 2\Omega(v) + 1) & \text{otherwise} \end{cases}$$

where $\mathrm{Act}(v, \overline{a}) = \{i \; : \; \Omega(v) \in A_i \wedge a_i = 2\}$ are the indices of active sets to which v belongs, p_i is the first component of the i-th memory $m_i = (p_i, c_i, q_i)$,

$p_{\max} = \max\{p_1 \ldots p_k\}$ and for each $A_i \in \mathcal{F}_0$ we have $r_i = |\{A_j \in \mathcal{F}_0 : A_i \subseteq A_j\}|$.

We have to prove that any play $\pi = v_0 v_1 v_2 \ldots$ of $\mathcal{G}$ is won by the same player as the extended play

$$\pi' = (v_0, m_{10}, \ldots, m_{k0}, a_{10}, \ldots a_{k0})(v_1, m_{11}, \ldots, m_{k1}, a_{11}, \ldots a_{k1}) \ldots.$$

For a given play π of $\mathcal{G}$, we divide the sets $A_1, \ldots, A_k \in \mathcal{F}_0$ into three classes.

The good: A_i is a good set if A_i is active ($a_i = 2$) only finitely often in π.

The bad: A_i is a bad set, if $A_i \subseteq \mathrm{Inf}(\pi)$ and A_i is not a good set.

The ugly: A_i is an ugly set if there is a priority $c \in A_i \setminus \mathrm{Inf}(\pi)$ and A_i is not a good set.

Lemma 4.21. *If A_i is bad and A_j is ugly, then $A_i \subseteq A_j$.*

Proof. Assume that there is a $b \in A_i \setminus A_j$. Since $A_i \subseteq \mathrm{Inf}(\pi)$ the visiting sequence for A_i is updated infinitely often, hence infinitely often with b, and whenever this happens then a_j is reset to 0. By definition there is a $c \in A_j$ that is seen only finitely many times in π. Therefore $a_j = 0$ from some point onwards. But this contradicts the assumption that A_j is not good. $\square$

We first assume that $\mathrm{Inf}(\pi) = A_i$ and prove that either no priority at all occurs infinitely often in π' or the minimal such priority is even.

Since from some point on there is no priority $d \notin A_i$ that occurs infinitely often, then for all sets A_j that are not subsets of A_i the visiting sequence will not be updated any more, and so the sequence $(p_{jn})_{n \in \omega}$ stabilizes at some value p_j. Since the visiting sequence of A_i is updated infinitely often, we get that from some point on $a_i = 2$. Hence A_i is a bad set. We can now argue as in the proof of Theorem 4.16: if infinitely many priorities appear in π, then the sequence $(p_{in})_{n \in \omega}$ diverges and no priority at all will be seen infinitely often in π'. It remains to consider the case where only finitely many priorities occur in π. Then the sequence $(p_{in})_{n \in \omega}$ stabilizes at some value p, which is the maximal priority appearing in π. For any $A_j \subsetneq A_i$, the sequence $(p_{jn})_{n \in \omega}$ will then also stabilize at the same value p, and $r_j > r_i$. It follows that some priority of form $2kp + 2r_\ell + 2$ occurs infinitely often in π', where $r_\ell \geq r_i$.

Suppose now that some smaller odd priority occurs infinitely often in π'. Then it would have to be of the form $2kp + 2r_j + 3$ with $r_j < r_\ell$ such that $a_j = 2$ infinitely often. However, only finitely many priorities appear in π. Hence if there are infinitely many positions v such that $\Omega(v) \in A_j$ and $a_j = 2$, then from some point onwards all these positions v satisfy that $\Omega(v) \in A_j \cap A_i$ and $a_i = 2$. On infinitely many such positions an update happens, and therefore, also the priority $2kp + 2r_j + 2$ appears infinitely often. Hence Player 0 wins π'.

For the converse, we now assume that Player 1 wins π.

Lemma 4.22. *Suppose that some even priority $2kq + 2r + 2$ is seen infinitely often in π'. Then q is the maximal priority that occurs in π and $r = r_\ell$ for some bad set A_ℓ.*

Proof. If there are infinitely many occurrences of $2kq + 2r + 2$ in π', then q is the maximal priority that occurs in π and some A_i is updated infinitely often (i.e. $A_i \subseteq \mathrm{Inf}(\pi)$) and active infinitely often. Obviously A_i is bad and $r \geq r_i$. If $r \neq r_\ell$ for all bad set A_ℓ, then $r = r_j$ for some other A_j that is active infinitely often. Thus A_j has to be ugly. But then by Lemma 4.21 $A_i \subseteq A_j$ and thus $r_i > r_j = r$. But $r \geq r_i$. $\qquad\square$

Let $r = \min\{r_\ell : A_\ell \text{ is bad}\}$. To show that Player 1 wins π' it suffices to prove that there is an odd priority occurring infinitely often in π' which, in case there exists a bound q on all priorities appearing in π, is smaller than $2kq + 2r + 2$.

Notice that for any ugly set A_i, the sequence $(p_{in})_{n \in \omega}$ stabilizes at some value p_i. Let $p = \max\{p_i : A_i \text{ is ugly}\}$.

We distinguish two cases. First we assume that there exists some priority

$$b \in \mathrm{Inf}(\pi) \setminus \bigcup\{A_i : A_i \text{ is bad}\}.$$

Fix n_0 such that, for all $n > n_0$, $p_{in} = p_i$ for all ugly sets A_i and $a_{in} \neq 2$ for all good sets A_i. Since $b \in \mathrm{Inf}(\pi)$ there exist infinitely many v_n with $n > n_0$ and $\Omega(v_n) = b$. For such v_n we have $\Omega'(v_n, \bar{m}_n, \bar{a}_n) = 2kp + 2r_i + 3$ if there is a set A_i (which has to be ugly) such that $b \in A_i$ and $a_i = 2$.

Otherwise $\Omega'(v_n, \bar{m}_n, \bar{a}_n)$ is odd and $\leq 2b + 1$. Since A_i is ugly and A_ℓ is bad it follows that $A_\ell \subseteq A_i$. Thus, $r_i < r$. Further $p \leq q$. It follows that there exists some odd priority $s \leq \max(2kp + 2r_i + 3, 2b + 1) < 2kq + r + 2$ that appears in π' infinitely often.

Now we consider the other case: every $b \in \mathrm{Inf}(\pi)$ is contained in some bad set $A_{i(b)}$. Let $A_1, \ldots, A_\ell$ be the bad sets. Without loss of generality, we assume that A_1 is a maximal bad set, i.e. $A_1 \not\subseteq A_i$ for $i = 2, \ldots, \ell$. Since A_1 is a strict subset of $\mathrm{Inf}(\pi)$, we can fix a priority $d \in \mathrm{Inf}(\pi) \setminus A_1$. Since any priority $d \in \mathrm{Inf}(\pi)$ is contained in some bad set, we can assume that $d \in A_2$. Further, by the maximality of A_1, we can fix priorities $e_2, \ldots, e_\ell$ where $e_i \in A_1 \setminus A_i$.

We consider a suffix of π that starts at a position where

- all sequences $(p_{in})_{n \in \omega}$ that stabilize at some value p_i have already reached that value,
- all good sets A_i have become inactive for good (i.e. $a_i \neq 2$),
- in the visiting sequence for A_1 the prefixes $p(e_1), \ldots p(e_\ell)$ have already been completed.

Note that A_1 is updated infinitely often, and between any two consecutive points in this suffix at which $\mathrm{up}_1 = 2$ all priorities $e_2, \ldots, e_\ell$ are seen at least once. Since the priority d appears infinitely often in π and A_2 is updated infinitely often, we are going to see infinitely many points v_{n_0} in the considered suffix of π for which $\Omega(v_{n_0}) = d$ and $a_1 = 0$ (since a_1 is reset with an update of A_2). Since a_1 increases to 2 infinitely often, there are infinitely many tuples $n_0 < n_1 < n_2$ such that $a_1 = i$ at all positions v_n with $n_i \leq n < n_{i+1}$ and $a_1 = 2$ at v_{n_2}.

By definition $\mathrm{up}_1 = 2$ at v_{n_1} and v_{n_2} and there cannot be any updates on priority d between v_{n_1} and v_{n_2}, as then a_1 would be reset to 0. By our choice of the considered suffix of π, there are updates on all $e_2, \ldots, e_\ell$ between v_{n_1} and v_{n_2}. Therefore, for any bad set A_j that contains d, we have that $a_j < 2$ between position v_{n_2} and the first position v_n with $\Omega(v_n) = d$ that comes after v_{n_2}. This is the case because between v_{n_1} and v_{n_2} the value a_j was reset to 0 by the update of the visiting sequence for A_1 by priority e_j, and since then it has not increased by more than 1 since there was no update on priority d.

Let us now consider the new priority at v_n. Since all bad sets A_j containing d are inactive, we have the same situation as in the first case: $\Omega'(v_n, \bar{m}_n, \bar{a}_n) = 2kp + 2r_i + 3$ if there is a set A_i (which has to be ugly) such that $d \in A_i$ and $a_i = 2$. Otherwise $\Omega'(v_n, \bar{m}_n, \bar{a}_n)$ is odd and $\leq 2d + 1$. Since A_i is ugly and A_ℓ is bad it follows that $A_\ell \subseteq A_i$ and thus $r_i < r_\ell = r$. Further $p \leq q$.

There are infinitely many such positions v_n. Thus there must exist some odd priority $s \leq \max(2kp + 2r_i + 3, 2d + 1) < 2kq + r + 2$ that appears in π' infinitely often. □

Of course, the same arguments apply to the case where $\mathcal{F}_1$ is finite.

Corollary 4.23. *Let $(\mathcal{F}_0, \mathcal{F}_1)$ be a Muller winning condition such that either $\mathcal{F}_0$ or $\mathcal{F}_1$ is finite. Then every Muller game with this winning condition is determined via FAR memory.*

4.5.4 Max-parity Games with Bounded Moves

We say that a an arena (G, Ω) has *bounded moves* if there is a natural number d such that $|\Omega(v) - \Omega(w)| \leq d$ for all moves (v, w) of G.

We have shown in Proposition 4.8 that, in general, winning strategies for max-parity games require infinite memory, but we do not know whether max-parity games are determined via FAR-memory.

For max-parity games with bounded moves, it is still the case that winning strategies may require infinite memory, but now we can prove determinacy via FAR-memory.

Proposition 4.24. *There exist max-parity games with bounded moves whose winning strategies require infinite memory.*

Proof. Consider a (solitaire) max-parity game with a single node v_0 of priority 0 from which Player 0 has, for every odd number $2n + 1$, the option to go through a cycle C_n consisting of nodes with priorities $2, 4, \ldots, 2n, 2n + 1, 2n, 2n - 2, \ldots, 4, 2$ and back to the node v_0. All these cycles intersect only at v_0. Clearly Player 0 has a winning strategy, namely to go successively through cycles $C_1, C_2, \ldots$ with the result that there is no maximal priority occurring infinitely often. However, if Player 0 moves according to a finite-memory strategy then only finitely many cycles will be visited and there is a maximal n such that the cycle C_n will be visited infinitely often. Thus the maximal priority seen infinitely often will be $2n + 1$ and Player 0 loses. □

Lemma 4.25. *Let π be a play of a max-parity game $\mathcal{G}$ with bounded moves such that infinitely many different priorities occur in π. Then $\max(\mathrm{Inf}(\pi))$ does not exist, so π is won by Player 0.*

Proof. Assume that moves of $\mathcal{G}$ are bounded by d and $\mathrm{Inf}(\pi) \neq \emptyset$ and let q be any priority occurring infinitely often on π. Since infinitely many different priorities occur on π it must happen infinitely often that from a position with priority q the play eventually reaches a priority larger than $q+d$. Since moves are bounded by d, this means that on the way the play has to go through at least one of the priorities $q+1, \ldots, q+d$. Hence at least one of these priorities also occurs infinitely often, so q cannot be maximal in $\mathrm{Inf}(\pi)$. □

Theorem 4.26. *Every max-parity game with bounded moves can be reduced via a one-dimensional* FAR-*memory to a parity game. Hence max-parity games are determined via strategies with one-dimensional* FAR-*memory.*

Proof. The FAR-memory simply stores the maximal priority m that has been seen so far. To reduce a max-parity game $\mathcal{G}$ with bounded moves, via this memory, to a parity game $\mathcal{G}'$ we define the priorities of $\mathcal{G}'$ by

$$\Omega'(v, m) = 2m - \Omega(v).$$

Let π be a play of $\mathcal{G}$ and let π' be the extended play in $\mathcal{G}'$. We distinguish two cases. First, we assume that on π the sequence of values for m is unbounded. This means that infinitely many different priorities occur on π, so by Lemma 4.25, Player 0 wins π. But since $m \leq \Omega'(v, m)$ and m never stabilizes there is no priority that occurs infinitely often on π, so π' is also won by Player 0.

In the second case there exists a suffix of π on which m remains fixed on the maximal priority of π. In that case $\mathrm{Inf}(\pi)$ is a non-empty subset of $\{0, \ldots, m\}$ and $\mathrm{Inf}(\pi')$ is a non-empty subset of $\{m, \ldots, 2m\}$. Further, $\Omega'(v, m)$ is even if, and only if, $\Omega(v)$ is even, and $\Omega'(v_1, m) < \Omega'(v_2, m)$ if, and only if, $\Omega(v_1) > \Omega(v_2)$. Thus, $\min(\mathrm{Inf}(\pi'))$ is even if, and only if, $\max(\mathrm{Inf}(\pi))$ is even. Hence π is won by the same Player as π'. □

4.6 Infinitary Zielonka-Tree Memory

In this section we stop investigating finite appearance records and study Muller conditions represented by Zielonka trees. These trees, with nodes labeled by sets of priorities, were introduced by Zielonka in [88] under the name of split trees to establish how much memory is needed for strategies in games with a fixed Muller condition.

Definition 4.27 (Cf. [88]). *The Zielonka tree for a Muller condition $(\mathcal{F}_0, \mathcal{F}_1)$ over a set C of priorities is a tree $Z(\mathcal{F}_0, \mathcal{F}_1)$ whose nodes are labeled with pairs (X, σ) such that $X \in \mathcal{F}_\sigma$. We define $Z(\mathcal{F}_0, \mathcal{F}_1)$ inductively as follows. Let*

$C \in \mathcal{F}_\sigma$ and $C_0, \ldots, C_k$ be the maximal sets in $\{X \subseteq C : X \in \mathcal{F}_{1-\sigma}\}$. Then $Z(\mathcal{F}_0, \mathcal{F}_1)$ consists of a root, labeled by (C, σ), to which we attach as subtrees the Zielonka trees $Z(\mathcal{F}_0 \cap \mathcal{P}(C_i), \mathcal{F}_1 \cap \mathcal{P}(C_i))$ for $i = 0, \ldots, k$. Moreover, if the intersection of all sets on an infinite branch of a Zielonka tree is not empty, then the intersection is added as the final point on the branch (so a Zielonka tree may be an ω-tree).

Every Muller condition over a finite set of priorities can be represented using a Zielonka tree, but this is not always possible for infinite sets C. This happens because there may be sets $D \in \mathcal{F}_\sigma$ that have subsets in $\mathcal{F}_{1-\sigma}$ but no maximal ones.

We will analyze Muller conditions $(\mathcal{F}_0, \mathcal{F}_1)$ over ω for which the tree $Z(\mathcal{F}_0, \mathcal{F}_1)$ exists, is finitely branching, and all internal nodes of the tree are labeled with co-finite sets. Slightly abusing notation, we will identify a node (X, σ) in the Zielonka tree with X, if the Muller condition is clear in the context. For such a node X, we denote its successors by $X^0, X^1, \ldots, X^k$ and, as all internal sets are co-finite, we know that $X \setminus X^i$ is finite for all i. Please note that with this notation we have implicitly fixed an ordering of the successors of each vertex.

We define the *height* of a node X in a Zielonka tree, denoted $h(X)$, as its distance to the root, starting with 0 if $\omega \in \mathcal{F}_0$ or with 1 if $\omega \in \mathcal{F}_1$. In this way the height of a node (X, σ) is even for $\sigma = 0$ and odd for $\sigma = 1$.

Example 4.28. The parity condition has a very simple Zielonka tree, namely just a Zielonka path

$$\omega \;\to\; \omega \setminus \{0\} \;\to\; \omega \setminus \{0, 1\} \;\to\; \omega \setminus \{0, 1, 2\} \;\to\; \cdots$$

and $h(X) = \min(X)$. There is no Zielonka tree for the max-parity condition since $\omega \in \mathcal{F}_0$ has no maximal subset in $\mathcal{F}_1$ as $\mathcal{F}_1$ is not closed under unions of chains.

To define a memory structure for Muller conditions that have Zielonka trees with the properties described above we slightly deviate from the previous definition of a memory structure. In the rest of this chapter we use memory structures with *move update functions*. We define this memory in the same way as in Definition 4.2, with the only exception that this time the memory update function, update : $M \times V \times V \to M$, depends on both the start and on the end position of a move, not only on the final position as in the previous definition. The inductive definition of the memory state is thus changed to $m(v_0) = \text{init}(v_0)$ and

$$m(v_0 \ldots v_i v_{i+1}) = \text{update}(m(v_0 \ldots v_i), v_i, v_{i+1}).$$

The notion of a reduction presented before generalizes to memory structures with move update functions in the following way. For a game graph $G = (V, V_0, V_1, E)$ and a memory structure $\mathfrak{M} = (M, \text{update}, \text{init})$, we again define $G \times \mathfrak{M} = (V \times M, V_0 \times M, V_1 \times M, E_{\text{update}})$, but

$$E_{\text{update}} = \{(v, m)(v', m') : (v, v') \in E \text{ and } m' = \text{update}(m, v, v')\}.$$

While this is a natural generalization and it preserves all the properties of memory reductions described previously, it does not allow to fully operate on moves because the priority function Ω is still defined on positions. For this reason, we define *games with move labeling* as games where the labeling function $\Omega : E \to C$ assigns priorities to moves instead of positions.

The notion of memory reduction applies to games with move labellings in the same way as to games with labellings of positions. Moreover, parity games with move labellings are again positionally determined. Remarkably, in the setting where multiple moves with different priorities are allowed between any two positions, even a stronger relationship between parity winning conditions and positional determinacy can be established. Not only are parity winning conditions in this setting the only positionally determined Muller winning conditions, in the same sense as in Theorem 4.9, but this statement holds for all prefix-independent winning conditions, even if these are not Muller conditions and over arbitrary sets of priorities [23]. This motivates us to define a Zielonka tree memory with move update function.

Definition 4.29. *For a Muller condition $(\mathcal{F}_0, \mathcal{F}_1)$ over ω with finitely branching Zielonka tree $\mathcal{Z} = Z(\mathcal{F}_0, \mathcal{F}_1)$ we define the infinitary Zielonka-tree memory as any memory structure*

$$\mathsf{ZTM}(Z(\mathcal{F}_0, \mathcal{F}_1)) = (Z, \text{update}, \text{init})$$

where Z are the nodes of $\mathcal{Z}$, $\text{init}(v) = \omega$ and the move update function satisfies the following constraint. If $\text{update}(X, v, w) = X'$ then the following conditions hold:

(1) $\Omega(w) \in X'$ and X' is a minimal node with this property, i.e. there is no successor Y of X' in $\mathcal{Z}$ for which $\Omega(w) \in Y$,

(2) if $\Omega(w) \in X$ then X' lies in the subtree $\mathcal{Z}|_X$,

(3) if $\Omega(w) \notin X$ then let Y be the minimal predecessor of X for which $\Omega(w) \in Y$ and let Y^i be the successor of Y for which $X \in \mathcal{Z}|_{Y_i}$;
in this case if Y has k successors then

$$X' \in \mathcal{Z}|_{Y^{i+1 \bmod k}}.$$

Intuitively, a Zielonka tree memory assigns to every play a corresponding walk on the Zielonka tree. At each step of this walk (except for the first one) if the play is in a position v then the walk is in a node $X \ni \Omega(v)$ that has no successors containing $\Omega(v)$, as guaranteed by Condition (1) above. If the play moves from v to w and the priority $\Omega(w)$ is already contained in the current position X then the walk moves down in $\mathcal{Z}|_X$ to any minimal position containing $\Omega(w)$, as guaranteed by Condition (2). If $\Omega(w)$ is not contained in the current position X then the walk first goes up to the minimal predecessor Y containing $\Omega(w)$ and then chooses the next successor of Y and moves down to any minimal position containing $\Omega(w)$ in the subtree corresponding to that next successor. This structure of the walk can be exploited to reduce games with winning conditions that have Zielonka trees with certain properties to parity games with move labeling as follows.

Theorem 4.30. *Every game $\mathcal{G}$ with a Muller winning condition $(\mathcal{F}_0, \mathcal{F}_1)$ over ω such that $Z(\mathcal{F}_0, \mathcal{F}_1)$ is finitely branching, all its internal nodes are co-finite, and the intersection of sets on every infinite branch belongs to $\mathcal{F}_0$, reduces via ZTM memory to a parity game with move labeling.*

Proof. The given Muller game $\mathcal{G}$ is reduced to a parity game $\mathcal{G}'$ with the move labeling function defined by

$$\Omega'((v, X), (w, X')) = h(Y),$$

where Y is again the minimal common ancestor of X and X', the same as in the definition above.

We prove that any play $\pi = v_0 v_1 \ldots$ of $\mathcal{G}$ is won by the same player as the extended play $\pi' = (v_0, X_0)(v_1, X_1) \ldots$ in $\mathcal{G}'$. Let us denote by Y_i the lowest common ancestor of X_i and X_{i+1}. This notation allows us to imagine the ZTM memory corresponding to π as a walk through the Zielonka tree,

$$X_0 \to Y_0 \to X_1 \to Y_1 \to X_2 \to \ldots.$$

We consider two cases.

Case 1: $\mathrm{Inf}(\Omega'(\pi')) = \emptyset$. In this case the walk through the Zielonka tree progresses downwards and there exists a unique infinite branch of $Z(\mathcal{F}_0, \mathcal{F}_1)$ to which infinitely many Y_i belong. The set of priorities seen infinitely often in π is then the intersection of the sets on this branch. Thus, by assumption, Player 0 wins π, and by definition the same player wins π'.

Case 2: there exists a minimal priority $m = \min \mathrm{Inf} \Omega'(\pi')$.
First, we claim that there is exactly one node Y of the Zielonka tree with $h(Y) = m$ that appears infinitely often in the sequence of nodes $Y_0 Y_1 \ldots$. Indeed, let $Y_i Y_{i+1} \ldots$ be the suffix of this sequence such that in π' there are no positions with priority smaller than m after step i. Let us assume that Y_k and Y_{k+l} are two different nodes with priority m that appear consecutively in $Y_i Y_{i+1} \ldots$. Since

$$Y_k \to X_{k+1} \to \ldots \to X_{k+l} \to Y_{k+l}$$

is a walk in the Zielonka tree connecting two different nodes with equal heights, there must be a node Y_m in this walk that is both an ancestor of Y_k and Y_{k+l}, but then $h(Y_m) < h(Y_k) = m$, which contradicts the way the suffix $Y_i Y_{i+1} \ldots$ was chosen.

Let us now look at the suffix $Y_i Y_{i+1} \ldots$ where $Y_i = Y$ and the only node with priority m in this suffix is Y. By definition of the update function this means that all priorities in $\Omega(v_i v_{i+1} \ldots)$ are contained in Y and therefore $\inf(\pi) \subseteq Y$.

Moreover, again by Condition (3) of Definition 4.29, when visiting Y the update function always chooses the next successor and then the first successor again. Since the tree $Z(\mathcal{F}_0, \mathcal{F}_1)$ is finitely branching, each successor Y^i is chosen infinitely often, and the subtree rooted at Y^i is left infinitely many times during the walk through the Zielonka tree that corresponds to π. By definition of the update function, this means that for each Y^i we infinitely often encounter priorities in $Y \setminus Y^i$ during the play π. As $Y \setminus Y^i$ is finite, this means that there is a priority $c \in Y \setminus Y^i$ encountered infinitely often and thus $\mathrm{Inf}(\pi) \not\subseteq Y^i$. Therefore the play π is won by the player σ for which $Y \in \mathcal{F}_\sigma$, which, by the definition of height for the Zielonka tree, is the same player who wins π' with priority m.   □

Note that in the theorem above we reduce a Muller game with labels on positions to a parity games with labels on moves. Since parity games with labels on moves are positionally determined, one can use the positional strategy from the parity game to obtain a strategy σ for the original Muller game that is a function depending only on the current position in the game v and on the node of the Zielonka tree X.

Since in a ZTM memory it holds that X is the minimal node with $\Omega(v) \in X$ (except for the first step which is irrelevant), we can encode the position X using only v and the current branch of the Zielonka tree. If we denote by $[\mathcal{Z}]$ the set of all branches, i.e. maximal paths through the Zielonka tree $\mathcal{Z}$, then we can modify the strategy σ to get a strategy σ' that depends only on the branches, instead of the positions in the tree. The following consequence follows for Zielonka trees where $[Z(\mathcal{F}_0, \mathcal{F}_1)]$ is finite.

Corollary 4.31. *Every game $\mathcal{G}$ with a Muller winning condition $(\mathcal{F}_0, \mathcal{F}_1)$ over ω such that $Z(\mathcal{F}_0, \mathcal{F}_1)$ has finitely many (possibly infinite) branches, all its internal nodes are co-finite, and the intersection of sets on every infinite branch belongs to $\mathcal{F}_0$, is determined with a finite memory of size $|[Z(\mathcal{F}_0, \mathcal{F}_1)]|$.*

Please note that this is both a generalization of positional determinacy for parity games with countably many priorities and of the classical result [88] that Muller games over a finite set of priorities are determined with finite memory equal in size to the number of leafs of $Z(\mathcal{F}_0, \mathcal{F}_1)$, which is of course the same as $|[Z(\mathcal{F}_0, \mathcal{F}_1)]|$ for finite trees.

5

Counting Quantifiers on Automatic Structures

In chapter 2 we asked how first-order logic can be extended and analyzed using infinitary logic, which lead us to the regular game quantifier and to clarifying the connection to games.

In this chapter we consider extensions of first-order logic in another direction, by generalized unary quantifiers. As usual, we require these extensions to preserve regularity on automatic structures. It turns out that the only generalized unary quantifiers with this property are the counting quantifiers:

- the *modulo counting quantifiers* "there exist k mod m many",
- the *infinity quantifier* "there exist infinitely many", and
- the *uncountability quantifier* "there exist uncountably many".

While it is known that all counting quantifiers indeed preserve regularity over finite-word automatic structures, and even over injectively presented ω-automatic structures, this was open for general ω-automatic structures. Our proof [10] uses ω-semigroups and leads to an additional corollary that all countable ω-automatic structures have injective presentations. It follows that countable ω-automatic structures have automatic presentations over finite words, which answers a question of Blumensath [11].

5.1 Generalized Quantifiers Preserving Regularity

To extend first-order logic with additional quantifiers it is useful to have an abstract definition of a *generalized quantifier*. We borrow the definition given by Lindström [60].

Definition 5.1. *A generalized quantifier Q over a relational signature $\tau = \{R_1, \ldots, R_k\}$ is a class of structures with signature τ that is closed under isomorphism. Let $\mathfrak{A}$ be a structure and $\varphi_1(\overline{x}_1, \overline{z}), \ldots, \varphi_k(\overline{x}_k, \overline{z})$ formulas over the signature $\sigma(\mathfrak{A})$ possibly different from τ, such that $|\overline{x}_i| = r_i$, i.e. the length of the vector $\overline{x}_i$ is the same as the arity of R_i. In first-order logic extended with the quantifier Q we allow to write formulas of the form $Q\overline{x}_1 \ldots \overline{x}_k(\varphi_1, \ldots, \varphi_k)$*

and define their semantics in the following way. If θ maps $\bar{z}$ to the tuple $\bar{a}$ of elements of $\mathfrak{A}$ then

$$\mathfrak{A}, \theta \models Q\bar{x}_1 \ldots \bar{x}_k(\varphi_1, \ldots, \varphi_k) \Leftrightarrow (A, \varphi_1^{\mathfrak{A}}(-, \bar{a}), \ldots, \varphi_k^{\mathfrak{A}}(-, \bar{a})) \in Q,$$

where by $\varphi_i^{\mathfrak{A}}(-, \bar{a})$ we denote the relation satisfied by exactly those tuples $\bar{b}$ for which $\varphi_i^{\mathfrak{A}}(\bar{b}, \bar{a})$ holds. The arity of the quantifier Q is the maximum of the lengths $|\bar{x}_i|$, so a unary quantifier is one where each of the vectors $\bar{x}_i$ is just a single variable.

To illustrate this definition observe that the classical quantifier $\exists$ is given by $\{(A, X) : X \neq \emptyset\}$ and $\forall$ is given by $\{(A, X) : X = A\}$. The quantifiers "there exist infinitely many" or "there exist k mod m many" can be represented in a similar way, but we give the standard definition.

The extension of first-order logic with counting quantifiers, denoted FO[C], allows to write all quantifiers of the following form:

- $\exists^{(r \bmod m)} x \; \varphi$ meaning that the number of x satisfying φ is finite and is congruent to r mod m,
- $\exists^{\infty} x \; \varphi$ meaning that there are infinitely many x satisfying φ,
- $\exists^{\leq \aleph_0} x \; \varphi$ and $\exists^{> \aleph_0} x \; \varphi$ meaning that the cardinality of the set of all x satisfying φ is countable, or uncountable, respectively.

The logic FO[C] has intimate relation to quantifiers that preserve regularity. To define this relation we first need to say that a generalized quantifier Q preserves (ω-)regularity if for every (ω-)automatic presentation $\mathfrak{d}, f$ of a structure $\mathfrak{A}$ every formula

$$\psi(\bar{z}) = Q\bar{x}_1 \ldots \bar{x}_k(\varphi_1(\bar{x}_1, \bar{z}) \ldots \varphi_k(\bar{x}_1, \bar{z}))$$

defines a relation $\psi^{\mathfrak{A}}$ that is (ω-)regular in the presentation $\mathfrak{d}, f$, i.e. such that $f^{-1}(\psi^{\mathfrak{A}})$ is (ω-)regular.

Moreover, we say that a quantifier Q over a signature τ is definable in FO[C] (or any other extension of FO) if there exists a formula $\varphi_Q \in$ FO[C] over the same signature τ such that $Q = \{\mathfrak{A} : \mathfrak{A} \models \varphi\}$. We can now state the result that shows the relationship between FO[C] and regularity-preserving quantifiers: every unary quantifier that preserves (ω-)regularity is definable in FO[C]. This result is proved in [79] for regularity preserving quantifiers and the proof extends to ω-regularity preserving ones.

The remaining question is whether every quantifier in FO[C] preserves regularity, and whether it does so in an effective way. For finite-word automatic structures the basic Theorem 1.6 can be extended to FO[C] as follows.

Theorem 5.2 (Cf. [45, 53, 14]).

- There is an effective procedure that given an automatic presentation $\mathfrak{d}, f$ of a structure $\mathfrak{A}$, and given an FO[C] formula $\varphi(\bar{x})$ defining a k-ary relation R over $\mathfrak{A}$, constructs a k-tape synchronous automaton recognizing $f^{-1}(R)$.

- *The FO[C]-theory of every automatic structure is decidable.*
- *The class of automatic structures is closed under FO[C]-interpretations.*

It has been observed that Theorem 5.2 can be extended to *injective ω-automatic* presentations [55, 58]. Moreover, Kuske and Lohrey show that the cardinality of any set definable in FO[C] is either countable or equal to that of the continuum. In the next section we work to extend this result to all, not necessarily injective automatic structures.

5.2 Defining Uncountability Using Equal Ends

We characterize when there exist countably many words x satisfying a given formula *with parameters* $\varphi(x, \bar{z})$ in some ω-automatic structure $\mathfrak{A}$. The characterization is first-order expressible in an ω-automatic extension of $\mathfrak{A}$ by the equal ends relation $\sim_e$ and the quantifier rank of the resulting formula depends on a constant C, which itself depends on φ and on the given presentation of $\mathfrak{A}$.

Let us fix an ω-automatic presentation $\mathfrak{d}$ of a structure $\mathfrak{A}$ with congruence $\approx$, and a first-order formula $\varphi(x, \bar{z})$ in the language of $\mathfrak{A}$ with x and $\bar{z}$ as free variables.

Proposition 5.3. *There is a constant C, computable from the presentation $\mathfrak{d}$, so that for all tuples $\bar{z}$ of infinite words the following are equivalent:*

(i) *$\varphi(-, \bar{z})$ is satisfiable and $\approx$ restricted to the domain $\varphi(-, \bar{z})$ has countably many equivalence classes,*

(ii) *there exist C-many words $x_1, \ldots, x_C$, each satisfying $\varphi(-, \bar{z})$, so that every x satisfying $\varphi(-, \bar{z})$ is $\approx$-equivalent to some $y \sim_e x_i$; formally, the structure $(\mathfrak{A}, \approx, \sim_e)$ models the sentence*

$$\forall \bar{z} \left(\exists^{\leq \aleph_0} w \; \varphi(w, \bar{z}) \longleftrightarrow \right.$$

$$\exists x_1 \ldots x_C \left(\bigwedge_i \varphi(x_i, \bar{z}) \wedge \right.$$

$$\left. \left. \forall x \left(\varphi(x, \bar{z}) \to \exists y \left(y \approx x \wedge \bigvee_i y \sim_e x_i \right) \right) \right) \right).$$

Proof. Suppose $\mathfrak{d}$, $\mathfrak{A}$, and φ are given. Define C to be c^2, where c is the size of the largest ω-semigroup corresponding to any of the given automata from the presentation $\mathfrak{d}$ or corresponding to φ. We fix the parameters $\bar{z}$ and let $\approx$ denote the equivalence relation $\approx$ restricted to the domain of $\varphi(-, \bar{z})$.

(ii) $\Rightarrow$ (i): Condition *(ii)* and the fact that every $\sim_e$-class is countable imply that all words satisfying $\varphi(-, \bar{z})$ are contained in a countable number of $\approx$-classes.

$(i) \Rightarrow (ii)$: The negation of Condition (ii) says that given $D < C$ many words $x_1, \ldots, x_D$, each satisfying $\varphi(-, \bar{z})$, there exists a word x_{D+1} also satisfying $\varphi(-, \bar{z})$ whose $\approx$-class does not meet any of the $\sim_e$-classes of the x_i for $i \leq D$.

Thus we can inductively define words $x_1, \ldots, x_C$, each satisfying the formula $\varphi(-, \bar{z})$, and such that for $1 \leq i < j \leq C$ the $\approx$-class of x_j does not meet the $\sim_e$-class of x_i. In particular, the x_is are pairwise non-equivalent with respect to $\sim_e$.

The plan is to produce uncountably many pairwise non-$\approx$ words that satisfy $\varphi(-, \bar{z})$. In the first 'Ramsey step', similar to what is done in [58], we find two words from the given C many, say $x_1, x_2 \in \Sigma^*$, and a factorization $H \subseteq \mathbb{N}$ so that both words behave the same way along the factored sub-words with respect to the $\approx$- and φ-semigroups. In the second 'Coarsening step' we identify a technical property of finite semigroups recognizing transitive relations. This allows us to produce an altered factorization G and new, well-behaving words y_1, y_2. In the final step, the new words are 'shuffled along G' to produce continuum many pairwise non-$\approx$ words, each satisfying $\varphi(-, \bar{z})$.

5.2.1 Ramsey Step

This step effectively allows us to discard the parameters $\bar{z}$. Before we use Ramsey's theorem, we introduce a convenient notation to talk about factorizations of words.

Definition 5.4. *Let $A = a_1 < a_2 < \ldots$ be any infinite subset of $\mathbb{N}$ and $h : \Sigma^* \to S$ be a morphism into a finite semigroup S. For an ω-word $\alpha \in \Sigma^\omega$, and element $e \in S$, say that A is an h, e-homogeneous factorization of α if for all $n \in \mathbb{N}^+$, $h\big(\alpha[a_n, a_{n+1})\big) = e$.*

Observe the following facts.

- If A is an h, s-homogeneous factorization of α and $k \in \mathbb{N}^+$ then the set $\{a_{k \cdot i}\}_{i \in \mathbb{N}^+}$ is an h, s^k-homogeneous factorization of α.
- If A is an h, e-homogeneous factorization of α and e is idempotent, then every infinite $B \subset A$ is also an h, e-homogeneous factorization of α.

In the following we write w^φ and $w^\approx$ to denote the image of w under the semigroup morphism into the finite semigroup associated to φ and $\approx$, respectively, as determined by the presentation. Accordingly, we will speak of e.g. φ, s_i-homogeneous factorizations.

Let us now color every $\{n, m\} \in [\mathbb{N}]^2$ with $n < m$ by the tuple of ω-semigroup elements

$$\left(\left(\otimes (x_i, \bar{z})[n, m)^\varphi \right)_{0 \leq i \leq C} , \left(\otimes (x_i, x_j)[n, m)^\approx \right)_{0 \leq i \leq j \leq C} \right).$$

By Ramsey's theorem there exists an infinite $H \subset \mathbb{N}$ and a tuple of ω-semigroup elements

$$((s_i)_{1 \le i \le C}, (t_{(i,j)})_{1 \le i \le j \le C})$$

so that for all $0 \le i \le j \le C$,

- H is a φ, s_i-homogeneous factorization of the word $\otimes(x_i, \overline{z})$,
- H is a $\approx, t_{(i,j)}$-homogeneous factorization of the word $\otimes(x_i, x_j)$.

Note that by virtue of additivity of our coloring and Ramsey's theorem each of the s_i and $t_{(i,j)}$ above are idempotents. Since there are at most c-many s_is and c-many $t_{(i,i)}$s, there are at most c^2 many pairs $(s_i, t_{(i,i)})$ and so there must be two indices, we may suppose 1 and 2, with $s_1 = s_2$ and $t_{(1,1)} = t_{(2,2)}$.

5.2.2 Coarsening Step

For technical reasons we now refine H and alter x_1, x_2 so that the semigroup elements have certain additional properties.

To start with, using the fact that $x_1 \not\sim_e x_2$ and the facts we observed on homogeneous factorizations, we assume without loss of generality that H is coarse enough so that $x_1[h_n, h_{n+1}) \ne x_2[h_n, h_{n+1})$ for all $n \in \mathbb{N}$.

Lemma 5.5. *There exists a subset $G \subset H$, listed as $g_1 < g_2 < \ldots$, and ω-words y_1, y_2 with the following properties:*

(1) The words y_1 and y_2 are neither $\approx$-equivalent nor $\sim_e$-equivalent, and each satisfies $\varphi(-, \overline{z})$.

(2) There exists an idempotent φ-semigroup element s such that G is a φ, s-homogeneous factorization for each of $\otimes(y_1, \overline{z})$ and $\otimes(y_2, \overline{z})$.

(3) There exist idempotent $\approx$-semigroup elements $t, t^\uparrow, t^\downarrow$ so that for $y_j \in \{y_1, y_2\}$
- *both $t^\uparrow$ and $t^\downarrow$ absorb t*
- *$\otimes(y_j, y_j)[0, g_1)^\approx$ absorbs t*
- *G is an $\approx, t$-homogeneous factorization of $\otimes(y_j, y_j)$*
- *G is an $\approx, t^\uparrow$-homogeneous factorization of $\otimes(y_1, y_2)$*
- *G is an $\approx, t^\downarrow$-homogeneous factorization of $\otimes(y_2, y_1)$.*

Proof. Define ω-words $y_1 := x_2[0, h_2)x_1[h_2, \infty)$, and y_2 by

$$y_2[0, h_2) := x_2[0, h_2) \text{ and}$$
$$y_2[h_{2n}, h_{2n+2}) := x_2[h_{2n}, h_{2n+1})x_1[h_{2n+1}, h_{2n+2}) \text{ for } n > 0.$$

Item 1. Clearly, $y_1 \not\sim_e y_2$ and each $y_j \in \{y_1, y_2\}$ satisfies $\varphi(y_j, \overline{z})$ since by homogeneity and $s_1 = s_2$

$$\otimes(y_1, \overline{z})^\varphi = \otimes(x_2, \overline{z})[0, h_2)^\varphi s_1^\omega$$
$$= \otimes(x_2, \overline{z})[0, h_2)^\varphi s_2^\omega$$
$$= \otimes(x_2, \overline{z})^\varphi,$$

and similarly

$$\otimes(y_2, \bar{z})^\varphi = \otimes(x_2, \bar{z})[0, h_2)^\varphi (s_2 s_1)^\omega$$
$$= \otimes(x_2, \bar{z})[0, h_2)^\varphi s_2^\omega$$
$$= \otimes(x_2, \bar{z})^\varphi.$$

Next we check that $y_1 \not\approx y_2$.

$$\otimes(y_1, y_2)^\approx = \pi_\approx \Big(\otimes (x_2, x_2)[0, h_2)^\approx,$$
$$\big(\otimes (x_1, x_2)[h_{2n}, h_{2n+1})^\approx,$$
$$\otimes (x_1, x_1)[h_{2n+1}, h_{2n+2})^\approx \big)_{n \in \mathbb{N}^+} \Big)$$
$$= \otimes(x_2, x_2)[0, h_1)^\approx t_{(2,2)} \, (t_{(1,2)} t_{(1,1)})^\omega$$
$$= \otimes(x_2, x_2)[0, h_1)^\approx t_{(2,2)} t_{(2,2)} \, (t_{(1,2)} t_{(1,1)})^\omega$$
$$= \otimes(x_2, x_2)[0, h_1)^\approx t_{(2,2)} t_{(2,2)} \, (t_{(1,2)} t_{(2,2)})^\omega$$
$$= \otimes(x_2, x_2)[0, h_1)^\approx t_{(2,2)} \, (t_{(2,2)} t_{(1,2)})^\omega$$
$$= \pi_\approx \Big(\otimes (x_2, x_2)[0, h_2)^\approx,$$
$$\big(\otimes (x_2, x_2)[h_{2n}, h_{2n+1})^\approx,$$
$$\otimes (x_1, x_2)[h_{2n+1}, h_{2n+2})^\approx \big)_{n \in \mathbb{N}^+} \Big)$$
$$= \otimes(y_2, x_2)^\approx$$

Thus, if $y_1 \approx y_2$ then also $y_2 \approx x_2$ and so *by transitivity* $y_1 \approx x_2$. But since $y_1 \sim_e x_1$, the $\approx$-class of x_2 meets the $\sim_e$-class of x_1, contradicting the initial choice of the x_is.

Items 2 and 3. Define intermediate semigroup elements $q := s_1$, $r := t_{(1,1)}$, $r^\uparrow := t_{(1,2)} t_{(1,1)}$ and $r^\downarrow := t_{(2,1)} t_{(1,1)}$. Then

1. both $r^\uparrow$ and $r^\downarrow$ absorb r, since $t_{(1,1)}$ is idempotent,
2. $\otimes(y_j, y_j)[0, h_2)^\approx = \otimes(y_j, y_j)[0, h_1)^\approx t_{(2,2)}$ and thus absorbs r (for $y_j \in \{y_1, y_2\}$).

In this notation, for all $i \in \mathbb{N}^+$ and $y_j \in \{y_1, y_2\}$,

- $\otimes(y_j, \bar{z})[h_{2i}, h_{2i+2})^\varphi$ is $qq = q$,
- $\otimes(y_j, y_j)[h_{2i}, h_{2i+2})^\approx$ is $rr = r$,
- $\otimes(y_1, y_2)[h_{2i}, h_{2i+2})^\approx$ is $t_{(1,2)} t_{(1,1)} = r^\uparrow$,
- $\otimes(y_2, y_1)[h_{2i}, h_{2i+2})^\approx$ is $t_{(2,1)} t_{(1,1)} = r^\downarrow$.

Finally, define the set $G := \{h_{2ki}\}_{i>1}$, i.e. $g_i = h_{2k(i+1)}$, and the semigroup elements $t := r^k$, $t^\uparrow := (r^\uparrow)^k$, $t^\downarrow := (r^\downarrow)^k$ and $s := q^k$. The extra multiple of k (defined as the product of the exponents of the semigroups for $\sim_e$ and $\approx$) ensures all these semigroup elements (in particular $t^\uparrow$ and $t^\downarrow$) are idempotent. We now verify the absorption properties:

$$t^\uparrow t = r^{\uparrow k} r^k = r^{\uparrow k} = t^\uparrow \quad \text{because } r^\uparrow \text{ absorbs } r.$$

Similarly, $t^{\downarrow}t$ absorbs t. Further, since $g_1 = h_{4k}$, we have

$$\otimes(y_j, y_j)[0, g_1)^{\approx} = \otimes(y_j, y_j)[0, h_2)^{\approx} \otimes (y_j, y_j)[h_2, h_{4k})^{\approx}$$
$$= \otimes(y_j, y_j)[0, h_2)^{\approx} r^{4k-2}$$
$$= \otimes(y_j, y_j)[0, h_2)^{\approx} r^{3k-2} t$$

and thus absorbs t.

Finally, we verify the homogeneity properties. Observe that G is an $\approx, t^{\downarrow}$-homogeneous factorization of $\otimes(y_2, y_1)$ since for $i \in \mathbb{N}^+$

$$\otimes(y_2, y_1)[g_i, g_{i+1})^{\approx} = \otimes(y_2, y_1)[h_{2k(i+1)}, h_{2k(i+2)})^{\approx}$$
$$= (r^{\downarrow})^k = t^{\downarrow}.$$

The other cases are similar.   □

5.2.3 Shuffling Step

We continue the proof of Proposition 5.3 by shuffling the words y_1 and y_2 along G resulting in continuum many pairwise distinct words that are pairwise not $\approx$-equivalent, each satisfying $\varphi(-, \overline{z})$. To this end, we define for $S \subset \mathbb{N}^+$ the characteristic word χ_S by

$$\chi_S[0, g_1) := y_2[0, g_1) \text{ , and}$$
$$\chi_S[g_n, g_{n+1}) := \begin{cases} y_2[g_n, g_{n+1}) & \text{if } n \in S \\ y_1[g_n, g_{n+1}) & \text{otherwise} \end{cases}$$

First observe that $\mathfrak{A} \models \varphi(\chi_S, \overline{z})$. Indeed, by item (2) of Lemma 5.5

$$\otimes(\chi_S, \overline{z})^{\varphi} = \otimes(y_2, \overline{z})[0, g_1)^{\varphi} s^{\omega}$$
$$= \otimes(y_2, \overline{z})^{\varphi}$$

and $\mathfrak{A} \models \varphi(y_2, \overline{z})$ by item (1) of Lemma 5.5. Moreover, for $S \not\sim_e T$ the construction gives that $\chi_S \not\sim_e \chi_T$. This is due to our initial choice of $x_1 \not\sim_e x_2$ and the assumption that the factorization $(h_n)_{n \in \mathbb{N}}$ is coarse enough so that $x_1[h_n, h_{n+1}) \neq x_2[h_n, h_{n+1})$ and thus also $y_1[g_n, g_{n+1}) \neq y_2[g_n, g_{n+1})$ for all n.

The following two lemmas establish that if both $S \setminus T$ and $T \setminus S$ are infinite then $\chi_S \not\approx \chi_T$. We denote by $x_{\circ\bullet}$ the word $\chi_{2\mathbb{N}^+}$ and by $x_{\bullet\circ}$ the word $\chi_{2\mathbb{N}^+-1}$, and we write p for $\otimes(y_2, y_2)[0, g_1)^{\approx}$.

Lemma 5.6. *For all S, T such that both $S \setminus T$ and $T \setminus S$ are infinite*

$$\otimes(\chi_S, \chi_T)^{\approx} = \begin{cases} \otimes(x_{\circ\bullet}, x_{\bullet\circ})^{\approx} & \text{or} \\ \otimes(x_{\bullet\circ}, x_{\circ\bullet})^{\approx} \end{cases}$$

Proof. Define semigroup-elements p_n for $n \in \mathbb{N}$ by

$$
p_n := \begin{cases} t^{\downarrow} & \text{if } n \in S \setminus T \\ t^{\uparrow} & \text{if } n \in T \setminus S \\ t & \text{otherwise} \end{cases}
$$

Let m be the smallest number in $S \triangle T$. Suppose that $m \in S \setminus T$. Because both $t^{\uparrow}$ and $t^{\downarrow}$ are idempotent and since t is absorbed by both p, $t^{\uparrow}$ and $t^{\downarrow}$, and both $t^{\uparrow}$ and $t^{\downarrow}$ appear infinitely often (as both $S \setminus T$ and $T \setminus S$ are infinite), we have

$$
\otimes(\chi_S, \chi_T)^{\approx} = \pi_{\approx}(p, (p_n)_{n \in \mathbb{N}}) = p(t^{\downarrow}t^{\uparrow})^{\omega}
$$
$$
= \otimes(x_{\bullet\circ}, x_{\circ\bullet})^{\approx}.
$$

The case that $m \in T \setminus S$ similarly results in $\otimes(x_{\bullet\circ}, x_{\bullet\circ})^{\approx}$.   □

Lemma 5.7. $x_{\circ\bullet} \not\approx x_{\bullet\circ}$.

Proof. Define an intermediate word $x_{\circ\bullet\circ\circ} := \chi_{4\mathbb{N}^+ - 2}$. By computations similar to the above we find that

$$
\otimes(x_{\bullet\circ}, x_{\circ\bullet\circ\circ})^{\approx} = p(t^{\downarrow}t^{\uparrow}t^{\downarrow}t)^{\omega} = p(t^{\downarrow}t^{\uparrow}t^{\downarrow})^{\omega} = p(t^{\downarrow}t^{\uparrow})^{\omega}
$$
$$
= \otimes(x_{\bullet\circ}, x_{\circ\bullet})^{\approx}
$$

and

$$
\otimes(x_{\circ\bullet}, x_{\circ\bullet\circ\circ})^{\approx} = p(tttt^{\downarrow})^{\omega} = p(t^{\downarrow})^{\omega}
$$
$$
= \otimes(y_2, y_1)^{\approx}.
$$

Therefore, if $x_{\bullet\circ} \approx x_{\circ\bullet}$ then also $x_{\bullet\circ} \approx x_{\circ\bullet\circ\circ}$ and so by *symmetry and by transitivity* $x_{\circ\bullet} \approx x_{\circ\bullet\circ\circ}$. But in this case also $y_2 \approx y_1$, contradicting item (1) of Lemma 5.5.   □

We are now able to complete the proof of Proposition 5.3. There are continuum many classes in $\mathcal{P}(\mathbb{N})/\sim_{e}$, thus there is a continuum of pairwise non-$\sim_{e}$-equivalent sets S. To construct sets with pairwise infinite differences, we define for a set $S \subseteq \mathbb{N}$ the swap set

$$
\widehat{S} = \{2n + 1 \ : \ n \in S\} \cup \{2n + 2 \ : \ n \notin S\}.
$$

Observe that if $S \not\sim_{e} T$ then both $\widehat{S} \setminus \widehat{T}$ and $\widehat{T} \setminus \widehat{S}$ are infinite. Therefore taking the words $\chi_{\widehat{S}}$ for the continuum of pairwise non-$\sim_{e}$-equivalent sets S yields a continuum of non-$\approx$-equivalent words, each satisfying $\varphi(-, \bar{z})$.   □

5.3 FO[C] over ω-Automatic Structures

Using the results about countability of the previous section, we are finally able to extend Theorem 5.2 to ω-automatic structures.

Theorem 5.8. *The statements of Theorem 5.2 hold true for* FO[C] *over all (not necessarily injective) ω-automatic presentations.*

Proof. We prove the first item, i.e. give the procedure for constructing automata for formulas, from which the rest of the theorem follows immediately. We inductively eliminate occurrences of cardinality and modulo quantifiers in the following way.

The countability quantifier $\exists^{\leq \aleph_0}$ and uncountability quantifier $\exists^{> \aleph_0}$ can be eliminated (in an extension of the presentation by $\sim_e$) by the formula given in Proposition 5.3.

For the remaining quantifiers we further expand the presentation with the ω-regular relations

– $\pi(a, b, c)$ saying that $a \sim_e b \sim_e c$ and the last position where a differs from c is no larger than the last position where b differs from c, and
– $\lambda(a, b, c)$ saying that $\pi(a, b, c)$ and $\pi(b, a, c)$ and that the word $a[0, k]$ is lexicographically smaller than the word $b[0, k]$, where k is the common last position where a and b differ from c.

Now $\exists^{<\infty} x \, \varphi(x, \bar{z})$ is equivalent to

$$\exists x_1 \ldots x_C \, \Psi(x_1, \ldots, x_C, \bar{z})$$

where Ψ expresses that $x_1, \ldots x_C$ satisfy $\varphi(-, \bar{z})$ and there exists a position, say $k \in \mathbb{N}$, so that every $\approx$-class contains a word satisfying $\varphi(-, \bar{z})$ that coincides with one of the x_i from position k onwards. This additional condition can be expressed by

$$\exists y_1 \ldots y_C \forall x \exists y \left(\varphi(x, \bar{z}) \rightarrow x \approx y \wedge \bigvee_i \pi(y, y_i, x_i) \right).$$

Consequently, $\exists^{(r \bmod m)} x . \varphi(x, \bar{z})$ can be eliminated since we can pick out unique representatives of the $\approx$-classes. We write $i(w)$ for the smallest index i for which $w \sim_e x_i$. The representatives are those x that satisfy the following properties for every $y \neq x$ in the same $\approx$-class as x.

– Either the index $i(x) < i(y)$, or
– the index $i(x) = i(y)$ and $\lambda(x, y, x_{i(x)})$ holds.

Now we can apply the construction of [58] or [55] for elimination of the $\exists^{(r \bmod m)}$ quantifier. $\qquad \square$

5.4 Presentations of Countable ω-Automatic Structures

As a corollary of Proposition 5.3 we obtain that for every ω-regular equivalence relation with countably many classes a set of unique representatives is definable.

Corollary 5.9. *Let $\approx$ be an ω-automatic equivalence relation on Σ^ω. There is a constant C, depending on the presentation, so that the following are equivalent:*

(1) $\approx$ has countably many equivalence classes,
(2) there exist C many $\sim_e$-classes so that every $\approx$-class has a non-empty intersection with at least one of these $\sim_e$-classes.

If one of these conditions holds, then there exists an ω-regular set of representatives of $\approx$. Moreover, an automaton for this set can be effectively constructed given an automaton for $\approx$.

Proof. The first two items are simply a specialization of Proposition 5.3. We construct the ω-regular set of representatives as follows.

Write A for the domain of $\approx$ and consider the formula $\psi(x_1, \ldots, x_C)$ with free variables $x_1, \ldots, x_C$:

$$\bigwedge_i x_i \in A \;\wedge\; \forall x \in A \;\exists y\; (y \approx x \wedge \bigvee_i y \sim_e x_i)$$

The relation defined by ψ is ω-regular since it is a first order formula over ω-regular relations. By assumption it is non-empty, and therefore it contains an ultimately periodic word of the form $\otimes(a_1, \ldots, a_C)$. Each of these a_is is thus ultimately periodic, and we write $a_i = v_i(u_i)^\omega$.

By definition of ψ, every word has now an $\approx$-representative in $B = \bigcup_i \Sigma^*(u_i)^\omega$. It remains to prune B to select unique representatives for each $\approx$-class.

It is easy to construct an ω-regular well-founded linear order on B. For every $w \in B$, let $p(w) \in \Sigma^*$ be the length-lexicographically smallest word such that w has period $p(w)$. Also let $t(w) \in \Sigma^*$ be the length-lexicographically smallest word so that $w = t(w) \cdot p(w)^\omega$. Define an order $\prec$ on B by $w \prec w'$ if $p(w)$ is length-lexicographically smaller than $p(w')$, or otherwise if $p(w) = p(w')$ and $t(w)$ is length-lexicographically smaller than $t(w')$. The ordering $\prec$ is ω-regular since it is FO-definable in terms of ω-regular relations. Finally, the required set of representatives may be defined as the set of $\prec$-minimal elements of every $\approx$-class. An automaton for this set can be constructed from an automaton for $\approx$ as all the steps we made used definable relations. $\square$

Corollary 5.9 immediately yields an *injective* ω-automatic presentation from a given ω-automatic presentation. This is especially interesting together with the following proposition by which countable injective ω-automatic presentations can be transformed to automatic ones.

Proposition 5.10. *([11, Theorem 5.32]) Let $\mathfrak{d}$ be an injective ω-automatic presentation of a countable structure $\mathcal{A}$. Then, an (injective) automatic presentation $\mathfrak{d}'$ of $\mathcal{A}$ can be effectively constructed.*

Combining Proposition 5.10 and Corollary 5.9 we are able to answer affirmatively a question of Blumensath [11] and conclude that every countable ω-automatic structure is already automatic.

Corollary 5.11. *A countable structure is ω-automatic if and only if it is automatic. Transforming a presentation of one type into the other can be done effectively.*

The techniques used in this chapter not only give insight into the cardinality of the ω-automatic equivalence relations, but can also be used to study cliques built from an arbitrary binary ω-automatic relation. This was exploited recently in [57] to investigate which Ramsey-like theorems hold for ω-automatic structures.

Remarkably, the existence of injective presentations can not be extended from countable to arbitrary ω-automatic structures. Consider a disjoint sum of the Boolean algebra of sets of natural numbers $\mathfrak{B} = (\mathcal{P}(\mathbb{N}), \cup, \cap, ^C)$ and the uncountable atomless Boolean algebra $\mathfrak{B}/\sim_e$, where sets with finite symmetric difference are identified. Let us define the structure $\mathfrak{A} = (\mathfrak{B} \sqcup (\mathfrak{B}/\sim_e), B_1, B_2, f)$ which extends the disjoint sum $\mathfrak{B} \sqcup (\mathfrak{B}/\sim_e)$ with two predicates denoting the universes of the two components of the sum and a function that takes elements of $\mathfrak{B}$ to their corresponding classes in the other component of the disjoint sum, i.e. $f(B) = [B]_{\sim_e}$ for $B \in \mathfrak{B}$.

Observe that there is an ω-automatic presentation of $\mathfrak{A}$. Elements of $\mathfrak{A}$ are represented as ω-words over $\{0,1\}$ with the first bit indicating whether the word represents an element of $\mathfrak{B}$ or of $\mathfrak{B}/\sim_e$ and the other bits listing which numbers belong to the represented subset. Equality is defined as equality of words for words staring with 1, i.e. representing elements of $\mathfrak{B}$, and as $\sim_e$ for words starting with 0. Boolean operations can be represented by automata in the standard way and the function f must only check that the $\sim_e$-classes of the components coincide, which can be done by the $\sim_e$ automaton ignoring the first bit.

The fact that there is no injective ω-automatic presentation of the structure $\mathfrak{A}$ was recently shown by Hjörth, Khoussainov, Montalban and Nies [44]. The proof is based on the topological observation that certain morphism between $\mathfrak{B}/\sim_e$ and $\mathfrak{B}$ can not be Borel, which would be contradicted by an injective presentation of $\mathfrak{A}$. It follows that decidability of FO[C] on the structure $\mathfrak{A}$, which is a consequence of Theorem 5.8, can not be deduced from the previous Theorem 5.2, and so it shows that Theorem 5.8 is a strong generalization of the previous result.

6

Cardinality Quantifiers in MSO on Linear Orders

The intimate connection between first-order logic on automatic structures and MSO on $(\omega, <)$ can be used to define generalized-automatic structures, which we introduced in section 1.5 as the ones that are MSO-to-FO interpretable in a tree. It is therefore a natural extension of the previous work to ask whether counting quantifiers preserve regularity on such generalized-automatic structures.

In this and the next chapter we give a partial positive answer to this problem, as we investigate the cardinality quantifiers on injectively presented generalized-automatic structures. The restriction to injective presentations allows us to work directly with MSO and therefore, instead of extending first-order logic, we study the extension of MSO with the following quantifiers:

- the second-order infinity quantifier
  "there exist infinitely many sets X for which $\varphi(X)$ holds",
- and the second-order uncountability quantifier
  "there exist uncountably many sets X for which $\varphi(X)$ holds".

When working directly with MSO we do not use automata or semigroups as previously, but rely on the composition method instead, which allows us to consider linear orders and trees labeled with arbitrary predicates. First, we prove that on arbitrary countable structures the second-order infinity quantifier can be eliminated from MSO using the predicate that expresses infiniteness of a set, which is definable in MSO on finitely branching trees and linear orders. Further, we show that on arbitrary countable linear orders (this chapter) and trees (next chapter) the second-order uncountability quantifier can be eliminated as well, i.e. it can be expressed using pure MSO formulas. These results were obtained together with Vince Bárány and Alexander Rabinovich [7, 8, 9].

The main results of this chapter can be summarized as follows. Let us denote by $\mathrm{Unc}(X)$ the predicate meaning that the set X is uncountable, and let us say that a linear order is almost complete if its completion adds at most countably many points. We consider the extension of MSO with cardinality quantifiers $\exists^\kappa$, so that the meaning of a formula $\exists^\kappa X \varphi$ is that there are *at least* κ many sets X satisfying φ, for each cardinal $\kappa \in \{\aleph_0, \aleph_1, 2^{\aleph_0}\}$. We sometimes write $\exists^\infty$ instead of $\exists^{\aleph_0}$.

Theorem 6.1. *For every MSO formula $\varphi(X, \overline{Y})$ there exists an MSO(Unc) formula $\psi(\overline{Y})$ that is equivalent to $\exists^{\aleph_1} X \varphi(X, \overline{Y})$ over the class of almost complete linear orders.*

Theorem 6.1 immediately yields complete elimination of the uncountability quantifier over countable scattered chains. Next we will prove this for chains of order type of the rationals, which ultimately enables the extension to all countable chains, as summarized below.

Theorem 6.2 (Elimination of the uncountability quantifier)

(1) For every MSO($\exists^{\aleph_1}$) formula $\varphi(\overline{Y})$ there exists an MSO formula $\psi(\overline{Y})$ that is equivalent to $\varphi(\overline{Y})$ over the class of all ordinals.

(2) For every MSO($\exists^{\aleph_1}$) formula $\varphi(\overline{Y})$ there exists an MSO formula $\psi(\overline{Y})$ that is equivalent to $\varphi(\overline{Y})$ over the class of all countable linear orders.

Furthermore, in all these cases ψ is computable from φ.

In addition to the above, the reduction will show that over countable linear orders the quantifiers $\exists^{\aleph_1} X$ and $\exists^{2^{\aleph_0}} X$ are equivalent, i.e. that the continuum hypothesis holds for MSO-definable families of sets.

Theorem 6.3. *Let $\varphi(X, \overline{Y})$ be an arbitrary MSO formula. Then $\exists^{\aleph_1} X \varphi(X, \overline{Y})$ is equivalent to $\exists^{2^{\aleph_0}} X \varphi(X, \overline{Y})$ over all countable chains.*

All of the above trivially extend to cardinality quantifiers $\exists^{\aleph_0} \overline{X}$, $\exists^{\aleph_1} \overline{X}$ and $\exists^{2^{\aleph_0}} \overline{X}$ counting finite tuples of sets given that for any cardinal $\kappa \geq \aleph_0$

$$\exists^\kappa (X_0, X_1)\varphi \equiv \exists^\kappa X_0 \exists X_1 \varphi \lor \exists^\kappa X_1 \exists X_0 \varphi .$$

6.1 Infinity Quantifier

To start with, let us show how to eliminate the infinity quantifier $\exists^\infty$ from monadic second-order formulas over any structure where infiniteness of a set is expressible in MSO. This yields a uniform elimination of the infinity quantifier from MSO formulas over the binary tree and countable linear orders, as finiteness of a set is expressible over these structures. Over the binary tree, infiniteness of a set X is, by König's Lemma, equivalent to the existence of an infinite path every element of which is a prefix of some node in X. Over a countable linear order, by Ramsey's theorem, infiniteness of X is equivalent to the existence of a subset $Y \subseteq X$ of type ω or ω^*, i.e. a set with every element having a direct successor (or predecessor) and with at most one limit point.

Proposition 6.4. *Over all structures, the infinity quantifier $\exists^\infty$ can be effectively eliminated from monadic second-order formulas using the "set X is finite" predicate.*

To prove this proposition, we show how occurrences of the infinity quantifier can be eliminated from formulas inductively, according to the following claim.

Claim. An MSO formula $\varphi(X, \overline{Y})$ is satisfied on a structure $\mathfrak{A}$ for fixed parameters $\overline{Y}$ by finitely many sets X if and only if there is a finite set Z such that any two distinct sets both satisfying φ differ on Z, i.e.

$$\exists \text{ finite } Z \; \forall X_1 X_2 \left((\varphi(X_1, \overline{Y}) \wedge \varphi(X_2, \overline{Y}) \wedge X_1 \neq X_2) \rightarrow \right.$$
$$\left. \exists z \in Z((z \in X_1 \wedge z \notin X_2) \vee (z \in X_2 \wedge z \notin X_1)) \right).$$

Proof.
($\Rightarrow$) Let $X_1, \ldots, X_k$ be the sets that satisfy $\varphi(X_i, \overline{Y})$. For every pair of distinct sets X_i, X_j choose an element $z_{i,j}$ which belongs to X_i but not to X_j. Define Z as the set of all chosen elements.
($\Leftarrow$) We show by induction that if the cardinality of Z is k then there are at most 2^k sets X which satisfy $\varphi(X, \overline{Y})$. If $k = 0$ then there are no two distinct sets, thus at most one set satisfies φ. For the induction step, assume that the cardinality of Z is $k + 1$ and pick any element $z \in Z$. Observe that each pair of sets satisfying φ and including z has to differ on $Z \setminus \{z\}$, thus by the inductive assumption there are at most 2^k such sets. Analogously, any pair of sets satisfying φ and not including z has to differ on $Z \setminus \{z\}$, so there are at most 2^k such sets. In total there are at most $2 \cdot 2^k = 2^{k+1}$ sets that satisfy φ and differ on Z. □

Observe that the following converse of Proposition 6.4 holds as well. The predicate "the set X is infinite" can be defined using the $\exists^\infty$ quantifier, e.g. by saying that there exist infinitely many singletons in X. Moreover, if we have only the quantifier $\exists^{2^{\aleph_0}}$ we can define "X is infinite" by $\exists^{2^{\aleph_0}} Y \; Y \subseteq X$ and thus we can define $\exists^\infty$ as well.

6.2   Unique and Doubling Intervals

To eliminate a single occurrence of the uncountability quantifier from a formula $\exists^{\aleph_1} X \; \varphi(X, \overline{Y})$ we will make extensive use of the following notions for intervals.

Definition 6.5. *Let $\mathcal{L}$ be a labeled chain, $X, \overline{Y}$ subsets of $\mathcal{L}$ such that $\mathcal{L} \models \varphi(X, \overline{Y})$, and I an interval of $\mathcal{L}$.*

(1) We say that I is a U-interval for $\varphi, X, \overline{Y}$ whenever $X \cap I$ is the unique subset of its type on $\mathcal{L}|_I$. More precisely, if $\mathcal{L}|_I \models \forall Z \; \tau(Z, \overline{Y}) \rightarrow Z = X$, where $\tau = \mathrm{Tp}^n(\mathcal{L}|_I, X, \overline{Y})$ with n equal to the quantifier rank of φ.

(2) I is a D-interval *for φ, X, $\overline{Y}$ if it is not a* U-interval.
(3) I is an unsplittable D-interval *for φ, X, $\overline{Y}$ if it cannot be split into disjoint D-intervals.*

The name "U-interval" attests to the fact that the set X in question is *uniquely* determined by its type on a given interval, as opposed to "D-intervals" offering two (or more) distinct choices for X with the same type on the interval, thus (at least) *doubling* the total number of choices for X over the entire domain. Whenever $\varphi, X, \overline{Y}$ are clear from the context we will take the liberty of saying "I is an U-interval" instead of "I is U-interval for $\varphi, X, \overline{Y}$", and similarly for D-intervals and unsplittable D-intervals.

Note that all of these notions can be formalized in MSO. For example, there is an MSO formula $\mathrm{DINT}_\varphi(X, \overline{Y}, I)$ expressing that I is a D-interval for $\varphi, X, \overline{Y}$ and a formula $\mathrm{UNSP}_\varphi(X, \overline{Y}, I)$ such that $L \models \mathrm{UNSP}_\varphi(X, \overline{Y}, I)$ if and only if I is an unsplittable D-interval for $\varphi, X, \overline{Y}$.

Lemma 6.6. *If there is an infinite set of pairwise disjoint D-intervals for some X satisfying $\varphi(X, \overline{Y})$ then there are at least continuum many such X.*

Proof. If there are infinitely many disjoint D-intervals for some X satisfying $\varphi(X, \overline{Y})$ then there is an increasing or decreasing ω-sequence of disjoint D-intervals among them. The two cases being symmetric, assume we have an increasing ω-sequence of D-intervals. The chain $(\mathcal{L}, X, \overline{Y})$ can then be written as a sum $\sum_{i \leq \omega} (\mathcal{L}|_{L_i}, X, \overline{Y})$ such that for all $i < \omega$ the interval L_i contains a D-interval, hence L_i is itself a D-interval for X, and allowing L_ω to be empty. This means that for all $i \in \omega$, there is a subset $X'_i \subseteq L_i$ such that $X'_i \neq X_i \cap L_i$ and $\mathrm{Tp}^n(\mathcal{L}|_{L_i}, X, \overline{Y}) = \mathrm{Tp}^n(\mathcal{L}|_{L_i}, X'_i, \overline{Y})$, where n is the quantifier rank of φ. We associate to every $H \subseteq \omega$ the distinct set

$$X_H = \bigcup \{X'_i \mid i \in H\} \cup \bigcup \{X \cap L_i \mid i \notin H\}.$$

From Theorem 1.11 it follows that $\mathrm{Tp}^n(\mathcal{L}, X_H, \overline{Y}) = \mathrm{Tp}^n(\mathcal{L}, X, \overline{Y})$ for every $H \subseteq \omega$. In particular, each of the continuum many X_H satisfies $\varphi(X_H, \overline{Y})$ in $\mathcal{L}$.   □

This observation motivates the introduction of the following key notion.

Definition 6.7 (Finite U-U cover). *Let $(L, <, X, \overline{Y})$ be a labeled chain such that $L \models \varphi(X, \overline{Y})$ and I an interval of L. Intervals $I_1 \ldots I_k$ constitute a* finite U-U cover *of I for $\varphi, X, \overline{Y}$ if $I = \bigcup_j I_j$ and each I_j is either a U-interval or an unsplittable D-interval for $\varphi, X, \overline{Y}$.*

Again, we will most often take no mention of either φ, X, or $\overline{Y}$ when these are understood, or can be arbitrary. Let us start with an example illustrating the notions introduced above.

Example 6.8. Consider the formula $\varphi(X) = \forall xy \ (x < y \land y \in X \to x \in X)$ defining downwards-closed sets, i.e. cuts, on $L = (\mathbb{Q}, <)$. Observe that $\mathbb{Q}$

is itself an unsplittable D-interval constituting a trivial finite U-U cover for every cut X. For every proper cut X there is a unique finite U-U cover of $\mathbb{Q}$ consisting of two U-intervals, namely $(-\infty, \sup X)$ and $(\sup X, \infty)$. Let X_π be the set of all rationals smaller than π. Note that $[3, 4]$ is a D-interval for X_π because for any $r \in [3, 4] \setminus \mathbb{Q}$, the set $\{x \in \mathbb{Q} \mid x < r\}$ has the same type as X_π on $[3, 4]$. In $[3, 4]$ we can find a left and a right point, say $l = 3.1$ and $r = 3.5$, such that both $[l, 4]$ and $[3, r]$ are D-intervals for X_π – for this reason we will later say that the interval $[3, 4]$ is unbalanced. In fact, every interval containing π in its interior is a D-interval for X_π – a property, which we later formalize by defining essential points for X_π.

Lemma 6.9. *If I has no finite U-U cover, then I can be split into two D-intervals such that one of them has no finite U-U cover.*

Proof. Because I has no finite U-U cover, it is necessarily a D-interval, but not an unsplittable D-interval. It can thus be split into D-intervals I_1, I_2 with $I_1 \cap I_2 = \emptyset$ and $I_1 \cup I_2 = I$. Now if both I_1 and I_2 had a finite U-U cover, then this would yield a finite U-U cover of I. Therefore I_1 or I_2 has no finite U-U cover. $\qquad\square$

As a conclusion we obtain the following lemma.

Lemma 6.10. *The following dichotomy holds for each X:*

(1) either I contains infinitely many disjoint D-intervals for X, or
(2) I has a finite U-U cover for X.

Proof. Assume that I has a finite U-U cover. Then it has a finite U-U cover $I_1, \ldots, I_k$ such that $I_i \cap I_j = \emptyset$ for all $1 \le i < j \le k$. As each I_j is either a U-interval or an unsplittable D-interval it cannot contain two disjoint D-intervals. This gives an upper bound $k + (k - 1)$ on the size of any collection of pairwise disjoint D-intervals inside I: at most k D-intervals each contained properly in separate I_j's and at most $k - 1$ D-intervals intersecting two or more of the I_j's.

Conversely, if $I_0 = I$ has no finite U-U cover then, by Lemma 6.9, it can be split into disjoint D-intervals I_1 and J_0, with I_1 having no finite U-U cover. By induction we obtain D-intervals I_n and J_n for X such that $I_{n+1} = I_n \setminus J_n$ has no finite U-U cover for X. Then, for $m < n$, we have $J_m \cap J_n \subseteq J_m \cap I_n = \emptyset$. Hence we have found infinitely many pairwise disjoint D-intervals J_n for X. $\qquad\square$

Next we refine the notion of finite U-U covers as follows.

Definition 6.11 (Balanced cover)

(1) An unsplittable D-interval I is left balanced if $I_{<v} = \{w \in I \mid w < v\}$ is a U-interval for every $v \in I$.
(2) Similarly, an unsplittable D-interval I is right balanced if for every $v \in I$ the interval $I_{>v} = \{w \in I \mid w > v\}$ is a U-interval.

(3) An unsplittable D-interval I is balanced *if it is either left-balanced or right-balanced.*

(4) A finite U-U cover $I_1, \ldots, I_k$ is balanced *if for each $1 \leq j \leq k$, I_j is either a U-interval or a balanced unsplittable interval.*

Lemma 6.12. *An interval I has a finite U-U cover for X if and only if I has a balanced U-U cover for X.*

Proof. We show that every non-balanced unsplittable D-interval I for X can be split into two intervals L and R constituting a balanced U-U cover of I for X. (It is important to point out that the split depends on X.) This immediately implies the conclusion of the lemma.

Let I be a non-balanced unsplittable D-interval for X. Because I is not right-balanced, there is a point $v \in I$ such that $I_{>v}$ is a D-interval, consequently, $I_{<v}$ must be a U-interval, since I cannot be split into two D-intervals. The following set is therefore not empty.

$$L = \{v \in I \mid I_{<v} \text{ is a U-interval for } X\}$$

By definition, L is a downward-closed subinterval of I, and it is either a U-interval or a left-balanced unsplittable D-interval.

Let $R = I \setminus L$. Because I is not left-balanced, R cannot be empty. Then one of R or L is a U-interval. We have seen that if L is a D-interval then it is left-balanced. Similarly, we need to show that if R is a D-interval then it is right-balanced. Notice that $I_{<v}$ is a D-interval for X for every $v \in R$, otherwise v would be in L. Therefore, since I cannot be split into disjoint D-intervals for X, $I_{>v}$ is a U-interval for X for every $v \in R$. Which means precisely that R is right-balanced. □

Lemma 6.13. *There is a function $N(k, l, \varphi)$ such that for every chain $\mathcal{L} = (L, <, \overline{Y})$ if $\{I_1, \ldots, I_k\}$ is a balanced U-U cover of an interval I of $\mathcal{L}$ for each of $X_1, \ldots, X_n$ then $n \leq N(k, l, \varphi)$ or $X_i \cap I = X_j \cap I$ for some $i \neq j$.*

Proof. Let K be the number of $\mathrm{qr}(\varphi)$-types in $l + 1$ variables. Then, if J is a U-interval for $K + 1$ sets then two of these must realize the same type on J and hence have to coincide on J. Assume now that J is left-balanced for $2K + 1$ sets $X_1, \ldots, X_{2K+1}$, so for each $v \in J$ the interval $J_{<v}$ is a U-interval for each of these sets X_i. If for each pair X_i, X_j with $i \neq j$ there is a point $p_{i,j} \in J$ on which these two sets differ, then all the $2K + 1$ sets differ on the interval $J_{\leq p}$ with $p = \max\{p_{i,j} \mid i, j \leq 2K + 1\}$. Therefore there are at least $K + 1$ among them which are pairwise different on $J_{<p}$, which contradicts the fact that $J_{<p}$ is a U-interval for all of these. The case of right-balanced intervals is treated symmetrically.

Classify the sets X_i into 2^k classes according to which of the $I_1, \ldots, I_k$ are left- or right-balanced for each X_i (considering U-intervals, say, as left-balanced). By the above, no class can contain more than $(2K)^k$ sets pairwise different on I. Therefore $N(k, l, \varphi) = (4K)^k$ satisfies the claim. □

Combining Lemmas 6.6, 6.10, 6.12 and 6.13 we obtain the following criterion.

Proposition 6.14. *Let $\mathcal{L} = (L, <, \overline{Y})$ be an chain and $\varphi(X, \overline{Y})$ an MSO-formula and $\aleph_1 \leq \kappa \leq 2^{\aleph_0}$. Then*

$$\mathcal{L} \models \neg \exists^{\kappa} X \; \varphi(X, \overline{Y})$$

if and only if there exists a subset U of the completion of L such that $|U| < \kappa$, and for every X satisfying $\varphi(X, \overline{Y})$ there is a finite balanced U-U cover of $\mathcal{L}$ the end-points of which lie in U.

Every point of the completion of L can be represented by a cut – a subset of L. Hence a direct formalization of this criterion referring to the cardinality of a set of cuts would require a third-order predicate. So what have we gained? For one, from Proposition 6.14 it is immediate that over countable scattered linear orders, where the restriction on U becomes vacuous, $\exists^{\aleph_1}$ and $\exists^{2^{\aleph_0}}$ are equivalent and can be effectively eliminated, cf. Corollary 6.15. In [59] Kuske and Lohrey obtained similar results for ω using more intricate automata techniques.

   More generally, we observe that over almost complete linear orders the condition stated in Proposition 6.14 for $\kappa = \aleph_1$ can be formulated in MSO(Unc). This yields further elimination results for $\exists^{\aleph_1}$, in particular, over ordinal chains.

   Later in Section 6.4 we will show that in general U can be replaced by a subset of L and a *definable set of cuts*. This will allow us to reduce cardinality quantifiers over arbitrary sets to the weaker first-order cardinality quantifiers and the ostensibly simpler use of cardinality quantifiers applied solely to definable sets of cuts. Of course, the latter subsumes also the first-order cardinality quantifiers.

6.3   Almost Complete Linear Orders

After the preparations of the previous section we are ready to prove the first item of Theorem 6.2 concerning the collapse of MSO($\exists^{\aleph_1}$) to MSO over the class of all ordinals. This will be a corollary of the elimination step embodied in Theorem 6.1 and valid uniformly over all *almost complete* linear orders. Recall that these have at most countably many proper cuts, i.e. cuts without a maximal element.

Theorem 6.1. *To every MSO-formula $\varphi(X, \overline{Y})$ one can effectively associate an MSO(Unc)-formula $\psi(\overline{Y})$ that is equivalent to $\exists^{\aleph_1} X \; \varphi(X, \overline{Y})$ over the class of almost complete linear orders.*

Proof. We are going to show that over almost complete linear orders the condition stated in Proposition 6.14 can be formulated in MSO(Unc). First, let

U be as in Proposition 6.14 and let $V = U \cap L$. Note that $V \cup (\overline{L} \setminus L)$ is an over-approximation of U, which also fulfills the condition stated in Proposition 6.14.

Let M be a subset of L. Define an equivalence relation $\sim_M$ as follows: $x \sim_M y$ if $[x, y] \subseteq M$ or $[x, y]$ is disjoint from M. Note that for every M, the equivalence classes of $\sim_M$ are intervals of L and the following can be formalized in MSO:

(i) The number of $\sim_M$ classes is finite.
(ii) Each $\sim_M$ class is a U-interval or a balanced unsplittable D-interval.
(iii) Whenever an end-point of a $\sim_M$-class lies in L then it is contained in V.

Note also, that if $I_0, \ldots I_k$ are disjoint intervals that partition L, then there is an M such that the I_i are $\sim_M$-equivalence classes. Indeed if for $i < j$ the interval I_i precedes I_j then we can take as M the set $I_0 \cup I_2 \cup \cdots \cup I_{2\lfloor k/2 \rfloor}$.

Over almost complete linear orders the conditions of Proposition 6.14 can thus be formalized by an MSO(Unc) formula expressing that there is a countable set V such that for all X satisfying $\varphi(X, \overline{Y})$ there is a set M such that the $\sim_M$ classes constitute a finite balanced U-U cover for X and all end-points of $\sim_M$ classes fall in $V \cup (\overline{L} \setminus L)$. □

In cases where the uncountability predicate, equivalently, the first-order uncountability quantifier is MSO-definable the above technique can be used inductively to completely reduce MSO($\exists^{\aleph_1}$) to MSO.

Corollary 6.15

(1) For every MSO($\exists^{\aleph_1}$) formula $\varphi(\overline{Y})$ there exists an MSO formula $\psi(\overline{Y})$ that is equivalent to $\varphi(\overline{Y})$ over the class of countable scattered linear orders.
(2) For every MSO($\exists^{\aleph_1}$) formula $\varphi(\overline{Y})$ there exists an MSO formula $\psi'(\overline{Y})$ that is equivalent to $\varphi(\overline{Y})$ over the class of all ordinals.
(3) (Under CH) For every MSO($\exists^{\aleph_1}$) formula $\varphi(\overline{Y})$ there exists an MSO formula $\psi''(\overline{Y})$ that is equivalent to $\varphi(\overline{Y})$ over the reals.

Proof. In all three cases one eliminates successively all uncountability quantifiers from ψ from the inside out by an application of Theorem 6.1 followed by the elimination of the predicate $\text{Unc}(X)$.

Over countable linear orders the predicate Unc is vacuously always false, hence the first claim.

Gurevich [36] proved (assuming the continuum hypothesis) that the predicate "the set X is uncountable" is expressible in MSO over the reals, which proves the third claim.

It is well-known that "the set X is uncountable" is expressible in MSO over the class of all ordinals. Recall that a subset X of an ordinal α is uncountable if the order type of $(X, <)$ is greater than or equal to ω_1, i.e. if there is a subset $Y \subseteq X$ such that the cofinality of the order-type of $(Y, <)$ is strictly greater than ω. This is the case precisely if every subset $Z \subseteq Y$ such that the order-type of $(Z, <)$ is ω is bounded in Y. Formally, "the set X is uncountable" is equivalent over ordinals to

$$\exists Y \subseteq X \ \forall Z \subseteq Y \ \omega(Z) \rightarrow \exists y \in Y \forall z \in Z \ z < y$$

where $\omega(Z)$ expresses that the order type of $(Z, <)$ is ω, for instance by saying that Z is infinite and $[0, z) \cap Z$ is finite for every $z \in Z$. Finiteness can be expressed e.g. as shown in Proposition 6.4. □

Let us stress again that, by Proposition 6.14, $\exists^{\aleph_1}$ and $\exists^{2^{\aleph_0}}$ are equivalent over all countable scattered linear orders.

6.4 Reduction to Counting Cuts

In this section we show that the existence of κ many sets satisfying an MSO formula can be reduced to the existence of κ many cuts (downward closed sets) satisfying some MSO formula effectively obtainable from the prior one. Cuts are of course just representations of points of the completion of the underlying linear order. Hence we show that a single use of the second-order cardinality quantifier $\exists^{\kappa} X$ over a linear order L reduces to a single use of the corresponding first-order cardinality quantifier $\exists^{\kappa} x$ over its completion $\overline{L}$.

We say that a point of $\overline{L}$ is a *splitting point* of a given finite U-U cover if it is an end-point of an interval in this cover. When convenient, we may blur the distinction between a cut C of a linear order L and the corresponding point $\sup C$ of the completion $\overline{L}$.

Definition 6.16. *A cut C is an* essential cut *for X, equivalently, $\sup C \in \overline{L}$ is an* essential point *for X, if every interval I such that I intersects both C and its complement is a D-interval for X.*

Lemma 6.17

(1) If $\sup C$ is an essential point for X then it is a splitting point of every finite balanced U-U cover for X.

(2) If there is a finite balanced U-U cover for some X then there is also one whose non-essential splitting points belong to L.

Proof

(1) Assume indirectly that there is an interval I of some balanced U-U cover for X and points $v, w \in I$ such that $v \in C$ and $w \notin C$. Then (v, w) must be a U-interval for X because either $\{x \in I \mid x < w\}$ or $\{x \in I \mid v < x\}$ is a U-interval and (v, w) is contained in both.

(2) Consider wlog. a finite balanced U-U cover consisting of disjoint intervals bounded by consecutive elements $\sigma_1 < \sigma_2 < \ldots < \sigma_t$ of the completion $\overline{L}$. If some $\sigma_j \in \overline{L} \setminus L$ is not an essential cut for X then, by definition, there is a U-interval I such that $\inf I < \sigma_j < \sup I$. Hence there are points $v, w \in L$ such that $[v, w] \subset I$ is a U-interval for X, and wlog. $\sigma_{j-1} < v < \sigma_j < w < \sigma_{j+1}$. The sequence $\sigma_1 < \ldots < \sigma_{j-1} < v < w < \sigma_{j+1} < \ldots < \sigma_t$ of splitting points thus gives rise to a new balanced U-U cover for X with

fewer non-essential cut-points in $\overline{L} \setminus L$. Continuing this way in a finite number of refinement steps we arrive at a finite balanced U-U cover for X all of whose non-essential splitting points belong to L.   □

Now we can refine Proposition 6.14 as follows.

Theorem 6.18. *Let $\mathcal{L} = (L, <, \overline{Y})$ be a chain, $\varphi(X, \overline{Y})$ an MSO formula and $\aleph_1 \leq \kappa \leq 2^{\aleph_0}$. Then*

$$\mathcal{L} \models \neg \exists^{\kappa} X \; \varphi(X, \overline{Y})$$

holds if and only if

(i) there exists a set $V \subseteq L$ such that $|V| < \kappa$ and every X satisfying $\varphi(X, \overline{Y})$ has a finite balanced U-U cover with non-essential splitting points in V, and

(ii) the total number of essential cuts for all X satisfying $\varphi(X, \overline{Y})$ is strictly less than κ.

Note the slight advantage of the above over Proposition 6.14. Condition (i) can be formalized using the first-order cardinality quantifier and (ii), unlike the condition of Proposition 6.14, refers to the cardinality of a definable set of cuts, i.e., a definable subset of the completion of $\mathcal{L}$.

Proposition 6.19. *To every MSO-formula $\varphi(X, \overline{Y})$ one can effectively associate MSO-formulas $\alpha(V, \overline{Y})$ and $\beta(C, V, \overline{Y})$ such that for all $\aleph_0 \leq \kappa \leq 2^{\aleph_0}$ over all linear orders $\exists^{\kappa} X \; \varphi$ is equivalent to $\forall V \, (\alpha \rightarrow \exists^{\kappa} C(\text{cut}(C) \wedge \beta))$.*

Note that, in particular, each of the above formulas makes use of only a single occurrence of the respective $\exists^{\kappa}$, and only restricted to cuts.

Proof. It is straightforward to give an MSO formula $\text{ECUT}_{\varphi}(C, X, \overline{Y})$ expressing that C is an essential cut for X. Thus, much as in the proof of Theorem 6.1, we can construct the formula $\alpha(V, \overline{Y})$ to express that "every X satisfying $\varphi(X, \overline{Y})$ has a finite balanced U-U cover with non-essential splitting points in V". This amounts to condition (i) of Theorem 6.18 without the cardinality constraint. Let further

$$\beta(C, V, \overline{Y}) = \text{``max} C \in V\text{''} \; \vee \; \exists X \left(\varphi(X, \overline{Y}) \wedge \text{ECUT}_{\varphi}(C, X, \overline{Y}) \right),$$

where $\text{max} C \in V$ is a shorthand for $\exists x \in V \; \forall y (y \in C \leftrightarrow y \leq x)$. Then there are at least κ many cuts C satisfying $\beta(C, V, \overline{Y})$ precisely if $|V| \geq \kappa$ or if condition (ii) of Theorem 6.18 fails.   □

6.5   Rationals

In this section we show how the uncountability quantifier can be eliminated in monadic second-order logic over the rationals. It will also be apparent that over the rationals $\exists^{\aleph_1} X \; \varphi$ and $\exists^{2^{\aleph_0}} X \; \varphi$ are equivalent for any MSO formula φ. Our argument makes use of a Ramsey-like theorem for additive colorings of dense chains due to Shelah.

Definition 6.20

(1) A coloring of a chain C is a function $col : [C]^2 \to T$ where $[C]^2$ is the set of unordered pairs of distinct elements of C and T is a finite set – the set of colors.

(2) The coloring f is additive if, for every $x_1 < y_1 < z_1$ and $x_2 < y_2 < z_2$ in C, it holds that $col(x_1, y_1) = col(x_2, y_2)$ and $col(y_1, z_1) = col(y_2, z_2)$ implies $col(x_1, z_1) = col(x_2, z_2)$. In this case a partial operation $+$ is well defined on T: $t_1 + t_2 = t$ if there are $x < y < z$ such that $col(x, y) = t_1$, $col(y, z) = t_2$ and $col(x, z) = t$.

(3) A sub-chain $D \subseteq C$ is homogeneous for col if there exists some $t_0 \in T$ such that for every $x, y \in D$, $col(x, y) = t_0$.

Shelah [80, Theorem 1.3] proved the following remarkable theorem.

Theorem 6.21 (Ramsey theorem for additive colorings). *Let col :* $[C]^2 \to T$ *be an additive coloring of a dense chain C using a finite set of colors T. Then there exists a homogeneous sub-chain $D \subseteq C$ for col that is everywhere dense in some open interval I of C.*

Recall that a *proper cut* is a cut having no supremum in the underlying linear order. A cut is non-trivial if neither it nor its complement is empty.

Lemma 6.22. *Let $\psi(C, Y_1, \ldots, Y_M)$ be an MSO formula and $V_1, \ldots, V_M \subseteq \mathbb{Q}$. Then there are uncountably many — and in fact continuum many — Dedekind cuts C of $\mathbb{Q}$ satisfying $(\mathbb{Q}, <) \models \psi(C, \overline{V})$ if and only if there is a subset $D \subseteq \mathbb{Q}$ such that $(D, <)$ is dense and for every non-trivial proper cut C of $(D, <)$ the Dedekind cut $C' = \{q \in \mathbb{Q} \mid \exists p \in C : q < p\}$ of rationals satisfies $\psi(C', \overline{V})$.*

Proof. Only the necessity of the above condition requires consideration. To that end assume that there are uncountably many cuts satisfying ψ and say that two rationals q and q' are *close*, $q \asymp q'$, if $[\min(q, q'), \max(q, q')]$ contains only countably many cuts satisfying ψ; and *far* otherwise. This defines an equivalence relation, each equivalence class of which is an interval of the rationals. These intervals are naturally linearly ordered and form a dense ordering. Indeed, by assumption there are at least two classes and by definition no two classes can form adjacent intervals, for otherwise their union would have to be part of a single class. In other words between any two points far apart there must be a third, which is far form both of these.

We assign to every pair $[q]_\asymp < [q']_\asymp$ of $\asymp$-classes as its color the n-theory of the interval $L_{[q,q')} = \bigcup\{[p]_\asymp \mid [q]_\asymp \le [p]_\asymp < [q']_\asymp\}$, where n is the quantifier rank of ψ:

$$\nu([q]_\asymp, [q']_\asymp) = \mathrm{Tp}^n(L_{[q,q')}, <, Y_1 \cap L_{[q,q')}, \ldots, Y_M \cap L_{[q,q')}).$$

By composition, ν defines an additive binary coloring on $\mathbb{Q}/_\asymp$. Thus, Theorem 6.21 asserts that there is an open interval $\mathcal{I}$ of $\mathbb{Q}/_\asymp$ and a subset $\mathcal{O} \subset \mathcal{I}$, which is dense in $\mathcal{I}$ and is ν-homogeneous. In other words there exists an n-theory τ such that $\nu([q]_{\asymp}, [q']_{\asymp}) = \tau$ for all $[q]_{\asymp} < [q']_{\asymp}$ in $\mathcal{O}$. Let D be an arbitrary complete set of representatives of the $\asymp$-classes in $\mathcal{I}$. In particular $(D, <)$ is dense, countable and without endpoints. Let $D_0 = \{q \in D \mid [q]_{\asymp} \in \mathcal{O}\}$ and let $I = \bigcup \mathcal{I}$.

Consider now a non-trivial proper cut C of D and let $C' = \{q \in \mathbb{Q} \mid \exists p \in C : q < p\}$ as in the statement of this lemma. Because D_0 is dense in D there exist $\mathbb{Z}$-chains $\ldots < p_{-2} < p_{-1} < p_0 < p_1 < \ldots$ in $D_0 \cap C$ and $\ldots < q_{-2} < q_{-1} < q_0 < q_1 < \ldots$ in $D_0 \setminus C$ such that $C = \{d \in D \mid \exists z \; p_z \le d < p_{z+1}\}$ and, similarly, $D \setminus C = \{d \in D \mid \exists z \; q_z \le d < q_{z+1}\}$. In particular, there is no $d \in D$ such that $p_z < d < q_z$ for all $z \in \mathbb{Z}$, which also means that there is in fact no $q \in \mathbb{Q}$ such that $p_z < q < q_z$ for all $z \in \mathbb{Z}$. Therefore we have $I \cap C' = \{q \in \mathbb{Q} \mid \exists z \; p_z \le q < p_{z+1}\}$ and $I \setminus C = \{q \in \mathbb{Q} \mid \exists z \; q_z \le q < q_{z+1}\}$.

By composition and homogeneity of $\mathcal{O}$, the n-types

$$\mathrm{Tp}^n(I \cap C', <, Y_1 \cap I \cap C', \ldots, Y_M \cap I \cap C')$$

and

$$\mathrm{Tp}^n(I \setminus C', <, Y_1 \cap I \setminus C', \ldots, Y_M \cap I \setminus C')$$

are obtained as the $\mathbb{Z}$-fold product of τ with itself and as such are independent of the choice of C. By the composition theorem again it follows that either every C' as above satisfies $\psi(C', Y_1, \ldots, Y_M)$ or none does. The latter possibility can be immediately ruled out on the grounds that any two points of D are by definition far apart meaning that there must be uncountably many Dedekind cuts between them satisfying ψ of which at most countably many do not induce, equivalently, are not induced by a proper cut of D.   □

The "only if" condition of Lemma 6.22 is clearly MSO expressible. Combined with Proposition 6.19 it yields full and effective elimination of $\exists^{\aleph_1}$ over $(\mathbb{Q}, <)$.

Proposition 6.23 (Elimination of $\exists^{\aleph_1}$ over the rationals). *For every* MSO*-formula $\varphi(X, \overline{Y})$ one can compute an* MSO*-formula $\psi(\overline{Y})$ which is equivalent over the standard ordering of the rationals to both $\exists^{\aleph_1} X \; \varphi(X, \overline{Y})$ and $\exists^{2^{\aleph_0}} X \; \varphi(X, \overline{Y})$.*

Proof. The proof is by induction on the structure of the formula. To eliminate an inner-most occurrence of the uncountability quantifier one applies first Proposition 6.19, followed by an application of Lemma 6.22.   □

6.6 Sums of Linear Orders

In the following we will make use of a more informative statement on composition of types on sums of linear orders as formulated by Shelah. Recall that we denote by $H_{n,k}$ the set of Hintikka formulas of quantifier depth n having k free variables (see Lemma 1.10).

Theorem 6.24 (Composition on linear orders II, [80])
Let $\varphi(\overline{X})$ be an MSO*-formula in the signature of chains with l predicates, m free variables and quantifier rank n. Given the enumeration $\tau_1(\overline{X}), \ldots, \tau_k(\overline{X})$ of $H_{n,l+m}$, there exists an* MSO*-formula $\theta(Q_1, \ldots, Q_k)$, computable from the above, such that for every linear order $\mathfrak{I} = (I, <^I)$ and family $\{\mathcal{L}_i \mid i \in I\}$ of chains and subsets $V_1, \ldots, V_m$ of $\sum_{i \in \mathfrak{I}} \mathcal{L}_i$,*

$$\sum_{i \in \mathfrak{I}} \mathcal{L}_i \models \varphi(\overline{V}) \quad \Longleftrightarrow \quad \mathfrak{I} \models \theta(Q_1, \ldots, Q_k)$$

where the predicates $\overline{Q}$ form a partition of I induced by $\overline{V}$ as follows:

$$Q_r = \{i \in I \mid \mathrm{Tp}^n(\mathcal{L}_i, \overline{V}) = \tau_r\} \text{ for each } r \in \{1, \ldots, k\}.$$

Using this theorem we can formulate some general conditions allowing to reduce the problem of eliminating the uncountability quantifier $\exists^{\aleph_1}$ over sums of linear orders to eliminating $\exists^{\aleph_1}$ over the index structure as well as eliminating it uniformly over the summands.

Lemma 6.25. *Let $\varphi(X, \overline{Y})$ be an* MSO *formula of quantifier rank n, $\mathcal{L} = \sum_{i \in I} \mathcal{L}_i$ an ordered sum of chains, and $\overline{V}$ subsets of $\mathcal{L}$. Let the enumeration of n-types be given by $\tau_1(X, \overline{Y}), \ldots, \tau_k(X, \overline{Y})$ and let $\theta(\overline{T})$ be as in Theorem 6.24. Then there are uncountably many $U \subseteq \mathcal{L}$ satisfying $\mathcal{L} \models \varphi(U, \overline{V})$ if and only if one of the following conditions holds:*

(a) there is one such U having infinitely many disjoint D-intervals, or
(b) there is one such U and an index $i \in I$ so that $\mathcal{L}_i \models \exists^{\aleph_1} Z\, \tau_r(Z, \overline{V}|_{\mathcal{L}_i})$
    where τ_r is the n-type of $(U|_{\mathcal{L}_i}, \overline{V}|_{\mathcal{L}_i})$ on $\mathcal{L}_i$, or
(c) the set of those partitions $\overline{P}$ of I that are induced by $\overline{V}$ and some U satisfying $\varphi(U, \overline{V})$ is uncountable. This can be expressed by an MSO$(\exists^{\aleph_1})$*-formula over the index structure:*

$$(I, <) \models \exists^{\aleph_1} \overline{P} : \mathrm{Part}(\overline{P}) \wedge \bigwedge_{r=1}^{k} P_r \subseteq Q_r \wedge \theta(\overline{P})$$

where $\mathrm{Part}(\overline{P})$ states that $\overline{P}$ partition I and for each $r = 1 \ldots k$ the set $Q_r = \{i \in I \mid \mathcal{L}_i \models \exists X' \tau_r(X', \overline{V}|_{\mathcal{L}_i})\}$.

If moreover $\exists^{\aleph_1}$ is equivalent to $\exists^{2^{\aleph_0}}$ both over the index structure and on each of the summands then these two quantifiers are also equivalent over the sum.

Proof. Each of the three conditions is sufficient to yield uncountably many sets U satisfying φ. For (a) this was proven in Lemma 6.6 by the weaker form of the composition theorem. Similarly, for (b) this also follows directly already from the weaker composition theorem. Finally, for condition (c) this follows from the fact that, for every one of the uncountably many tuples $\overline{P}$ accounted for, there is a distinct set U inducing the type-predicates $\overline{P}$ and fulfilling $\varphi(U, \overline{V})$.

Conversely, if condition (c) fails then there are only countably many colorings of I with type predicates $\overline{P}$ induced by some U satisfying $\varphi(U, \overline{V})$. By failure of (a) for each of these type predicates we have for all but finitely many indices i that $i \in P_r$ implies that τ_r uniquely defines $U \cap L_i$ from $\overline{V}|_{L_i}$. Finally, if condition (b) fails too, then on each of the finitely many remaining intervals L_i there are also only countably many choices for $U \cap L_i$.

To see that the formalization of condition (c) provided above is sound note that by Theorem 6.24 every U satisfying $\varphi(U, \overline{V})$ induces, together with $\overline{V}$, a partition $\overline{P}$ satisfying the given formula. Conversely, each tuple $\overline{P}$ satisfying it fulfills all the following: It forms a partition, it is induced by some set U together with $\overline{V}$ as ensured by $\bigwedge_{r=1}^{k} P_r \subseteq Q_r$, and every U inducing it must satisfy $\varphi(U, \overline{V})$ thanks to $\theta(\overline{P})$. $\square$

A crucial point as to the applicability of the above claim is that it assumes a given factorization of a linear order as a sum. We introduce the notion of definable splitting to facilitate the use of the above technique on classes of linear orders over which an appropriate factorization is uniformly definable.

Definition 6.26 (Splitting). *Let $\mathcal{L} = (L, <, \dots)$ be a chain and $\theta(x, y)$ a formula with first-order variables x and y.*

(1) We call θ a splitting of $\mathcal{L}$ if $\{(a, b) \in L^2 \mid \mathcal{L} \models \theta(a, b)\}$ is an equivalence relation every class of which is an interval.
(2) For a splitting θ of $\mathcal{L}$ let $\sim_\theta^{\mathcal{L}}$ denote the equivalence relation defined by θ in $\mathcal{L}$, $I_{\mathcal{L}/\theta}$ the set of $\sim_\theta^{\mathcal{L}}$-classes and $\mathrm{Ind}_{\mathcal{L}/\theta} = (I_{\mathcal{L}/\theta}, <)$ the natural ordering of $I_{\mathcal{L}/\theta}$ according to representatives. We call $\mathrm{Ind}_{\mathcal{L}/\theta}$ the indexing order of $\mathcal{L}$ and $\mathcal{S}_{\mathcal{L}/\theta} = \{\mathcal{L}|_I \mid I \in I_{\mathcal{L}/\theta}\}$ the summand structures of $\mathcal{L}$ w.r.t. θ.
(3) Let $\mathcal{C}$ be a class of labeled chains. We call θ a splitting of $\mathcal{C}$ if θ splits every $\mathcal{L} \in \mathcal{C}$. Let $\mathrm{Ind}_{\mathcal{C}/\theta} = \{\mathrm{Ind}_{\mathcal{L}/\theta} \mid \mathcal{L} \in \mathcal{C}\}$ and $\mathcal{S}_{\mathcal{C}/\theta} = \bigcup_{\mathcal{L} \in \mathcal{C}} \mathcal{S}_{\mathcal{L}/\theta}$. We call $\mathrm{Ind}_{\mathcal{C}/\theta}$ and $\mathcal{S}_{\mathcal{C}/\theta}$ the class of indexing chains of $\mathcal{C}$ and the class of summand structures of $\mathcal{C}$ w.r.t. θ, respectively.

Theorem 6.27. *Let $\mathcal{C}$ be a class of labeled chains and θ a splitting of $\mathcal{C}$. If*

(1) $\mathsf{MSO}(\exists^{\aleph_1})$ collapses effectively to MSO over the class of indexing chains of $\mathcal{C}$ w.r.t. θ, and
(2) $\mathsf{MSO}(\exists^{\aleph_1})$ collapses effectively to MSO over the class of summand structures of $\mathcal{C}$ w.r.t. θ,

then $\mathsf{MSO}(\exists^{\aleph_1})$ collapses effectively to MSO over $\mathcal{C}$.

Proof. Consider a formula $\varphi(X, \overline{Y})$ of MSO. We give a formula $\alpha \vee \beta \vee \gamma$ expressing in MSO the disjunction of the three conditions of Lemma 6.25 equivalent to $\exists^{\aleph_1} X \varphi(X, \overline{Y})$ uniformly over each $\mathcal{L} \in \mathcal{C}$ with the factorization as defined by θ. Let $\tau_1, \dots, \tau_k$ be an enumeration of $\mathrm{Tp}(n, 1+m)$ where n is the quantifier rank of φ and m the length of $\overline{Y}$.

Condition (a) can be expressed in MSO uniformly over all chains of a given signature, for instance by requiring the existence of an X satisfying $\varphi(X, \overline{Y})$

and an infinite set D such that every interval containing at least two points of D is a D-interval for X:

$$\alpha = \exists X \; \exists D \; \mathrm{Inf}(D) \wedge \forall \text{ interval } I \; (\exists d \neq d' \in D \cap I) \rightarrow \mathrm{DINT}_\varphi(X, \overline{Y}, I).$$

The use of $\mathrm{Inf}(D)$ above, meaning that the set D is infinite, is of course just a shorthand. It can be eliminated as in Proposition 6.4.

Condition (b) can be expressed in MSO relying on the elimination procedure for $\mathcal{S}_{\mathcal{C}/\theta}$. By the latter, one obtains for each n-type $\tau_r(X, \overline{Y})$ an MSO formula $\nu_r(\overline{Y})$ equivalent to $\exists^{\aleph_1} Z \; \tau_r(Z, \overline{Y})$ over $\mathcal{S}_{\mathcal{C}/\theta}$. Using these, condition (b) can be written as

$$\beta = \exists X \; \varphi(X, \overline{Y}) \wedge \exists I \; \exists x \; \mathrm{CLASS}_\theta(x, I) \wedge \bigvee_r (\tau_r^I(X, \overline{Y}) \wedge \nu_r^I(\overline{Y})),$$

where $\mathrm{CLASS}_\theta(x, I) = \forall y (y \in I \leftrightarrow \theta(x, y))$ defines I as the equivalence class of x with respect to θ, and where for a formula ψ we denote by ψ^I the relativization of ψ to I.

Finally, to express condition (c) of Lemma 6.25 one needs to

- choose a set I of representatives of all the equivalence classes defined by θ,
- relativize to I the MSO formula $\rho(\overline{Q})$ equivalent to $\exists^{\aleph_1} \overline{P} \ldots$ of condition (c) over $\mathit{Ind}_{\mathcal{C}/\theta}$ as delivered by the elimination procedure for this class,
- and substitute into ρ^I the sets $\overline{Q}$ defined as

$$\Omega(\overline{Q}) = \bigwedge_{r=1}^{k} \forall x \big(x \in Q_r \leftrightarrow \exists X', L \big(\mathrm{CLASS}_\theta(x, L) \wedge X' \subseteq L \wedge \tau_r^L(X', \overline{Y}) \big) \big).$$

With the customary shorthand "$\exists! y$" meaning "there is a unique y such that" this formula takes the form

$$\gamma = \exists I \; (\forall x \; \exists! y \in I \; \theta(x, y)) \wedge \exists Q_1, \ldots, Q_k \; \Omega(\overline{Q}) \wedge \rho^I(\overline{Q}). \qquad \square$$

6.7 All Countable Linear Orders

At last we are in a position to conclude that the quantifiers $\exists^{\aleph_1}$ and $\exists^{2^{\aleph_0}}$ are equivalent and can be effectively eliminated from $\mathsf{MSO}(\exists^{\aleph_1}, \exists^{2^{\aleph_0}})$ uniformly over all countable chains. Indeed, as recalled in the first chapter, every countable linear order arises as a dense sum of scattered linear orders, i.e. in the form $\sum_{q \in D} L_q$ where each L_q is a countable scattered linear order and D is either a singleton or is isomorphic to the standard ordering of the rationals with or without an additional minimal or maximal element.

Let $\theta(x, y)$ be the MSO-formula expressing that for no subset A of $L_{[x,y]}$ the order $(A, <)$ is dense, i.e. that $L_{[x,y]}$ is a scattered linear order. Over any countable chain, θ defines an equivalence relation partitioning it into intervals coinciding with the summands L_q as above. Thus, θ is a splitting of the class of all countable linear orders.

Taking advantage of Theorem 6.27 and using the previously proven collapse results over the class of countable scattered linear orders (Corollary 6.15) and over the rationals (Proposition 6.23), — which trivially extends to the rationals with either one or both endpoints added — we obtain uniform effective elimination of $\exists^{\aleph_1}$ over the class of all countable chains. This completes the proof of Theorem 6.2(2) and similarly Theorem 6.3.

Cardinality Quantifiers in MSO on Trees

In this chapter, we extend the results on second-order cardinality quantifiers, shown for linear orders in the previous chapter, to trees. Our main result, obtained together with Vince Bárány and Alexander Rabinovich [8, 9], is that the uncountability quantifier can be eliminated from MSO over trees.

Theorem 7.1. *For every* $\mathsf{MSO}(\exists^{\aleph_0}, \exists^{\aleph_1}, \exists^{2^{\aleph_0}})$ *formula* $\varphi(\overline{Y})$ *there exists an* MSO *formula* $\psi(\overline{Y})$, *computable from* φ, *that is equivalent to* $\varphi(\overline{Y})$ *over trees.*

In addition to the above, the reduction will show that over trees the quantifiers $\exists^{\aleph_1} X$ and $\exists^{2^{\aleph_0}} X$ are equivalent, i.e. that the continuum hypothesis holds for MSO-definable families of sets. Though not surprising, this is not obvious for it is known that in MSO one can define non-analytic classes of sets [70] and that the continuum hypothesis is independent of ZFC already for co-analytic sets [67].

Theorem 7.2. *On trees* $\exists^{\aleph_1} X \, \varphi(X, \overline{Y})$ *is equivalent to* $\exists^{2^{\aleph_0}} X \, \varphi(X, \overline{Y})$ *for every* MSO *formula* $\varphi(X, \overline{Y})$.

Our theorems translate to generalized-automatic structures, as formulated in the corollary below. They also supersede the previously mentioned results from [59] and generalize the theorem of Niwiński [69], which states that over the full binary tree the validity of $\exists^{\aleph_1} \overline{X} \, \varphi(\overline{X})$ is decidable and equivalent to that of $\exists^{2^{\aleph_0}} \overline{X} \, \varphi(\overline{X})$ for every MSO-formula $\varphi(\overline{X})$.

Corollary 7.3. *Every expansion of an injectively generalized-automatic structure by a relation definable in first-order logic with (first-order) cardinality quantifiers is also an injectively generalized-automatic structure.*

7.1 D-Nodes versus U-Nodes and Relevant Branches

To eliminate the uncountability quantifier over trees, we will again define suitable notions of U-nodes and D-nodes, similar to U-intervals and D-intervals used in the previous chapter. As our main tool, we will again use the composition method, in the form of the following theorem.

Theorem 7.4 (Composition Theorem for Trees II)

Let $\varphi(\overline{X})$ be an MSO-formula in the signature of trees with l predicates, having m free variables and quantifier rank n. Given the enumeration $\tau_1(\overline{X}), \ldots, \tau_k(\overline{X})$ of $H_{n,l+m}$, there exists an MSO-formula $\theta(Q_1, \ldots, Q_k)$ computable from φ such that for every tree $\mathfrak{I} = (I, <^I)$ and family $\{\mathfrak{I}_i \mid i \in I\}$ of trees and subsets $V_1, \ldots, V_m$ of $\sum_{i \in \mathfrak{I}} \mathfrak{I}_i$,

$$\sum_{i \in \mathfrak{I}} \mathfrak{I}_i \models \varphi(\overline{V}) \quad \Longleftrightarrow \quad \mathfrak{I} \models \theta(Q_1, \ldots, Q_k)$$

where $Q_r = \{i \in I \mid \mathrm{Tp}^n(\mathfrak{I}_i, \overline{V}) = \tau_r\}$ for each $r \in \{1, \ldots, k\}$.

A *tree segment*, or *interval*, of a tree is a connected and convex set I of nodes, i.e. such that for every $u, w \in I$ if u and w are incomparable, then their greatest common ancestor is in I, and if $u < w$ then for every $u < v < w$ also $v \in I$. Every tree segment has a minimal element and every subtree $\mathfrak{I}_z$ of a tree $\mathfrak{I}$ is a tree segment. More generally, the summands $\mathfrak{I}_i$ of any tree sum $\mathfrak{I} = \sum_{i \in \mathfrak{I}} \mathfrak{I}_i$ are tree segments of $\mathfrak{I}$. The terms 'interval' and 'tree segment' are used interchangeably.

We denote by $\mathfrak{I}|_I$ the restriction of a tree $\mathfrak{I}$ to the interval I. Alternatively, given a node z and a set Z of nodes of $\mathfrak{I}$ we use the notation $\mathfrak{I}_{z \backslash Z}$ for the restriction of $\mathfrak{I}$ to the tree segment $\mathfrak{I}_z \setminus (\bigcup_{w \in Z, z < w} \mathfrak{I}_w)$. Any interval I with a minimal element z can be written in the form $\mathfrak{I}_{z \backslash Z}$, where $Z = \{u \mid u \geq z \wedge u \notin I\}$. In particular, if B is a branch, $v, w \in B$ such that w is the immediate successor of v on B, then $T_{v \backslash B} = T_v \setminus T_w$. These notations are schematically depicted in Figure 7.1.

Fig. 7.1. A subtree $\mathfrak{I}_v$ and tree segments $\mathfrak{I}_{v \backslash \{u,w\}}$ and $\mathfrak{I}_{v \backslash B}$

Consider an MSO formula $\varphi(X, \overline{Y})$ over trees. To eliminate a single occurrence of the uncountability quantifier from $\exists^{\aleph_1} X \; \varphi(X, \overline{Y})$ over a tree $\mathfrak{I}$ we will make extensive use of the following notions for intervals. For the rest of this section we fix an MSO formula $\varphi(X, \overline{Y})$ over trees with l predicates and with $1 + m$ free variables — of which $\overline{Y} = (Y_1, \ldots, Y_m)$ will often be regarded as parameters — and of quantifier rank n.

Definition 7.5. *Let $\mathfrak{I}$ be a tree, $X, \overline{Y}$ subsets of $\mathfrak{I}$ such that $\mathfrak{I} \models \varphi(X, \overline{Y})$, and I an interval of $\mathfrak{I}$.*

(1) We say that I is a U-interval for φ, X, $\overline{Y}$ whenever $X \cap I$ is the unique subset of its type on $\mathfrak{T}|_I$. More precisely, if $\mathfrak{T}|_I \models \forall Z\, \tau(Z, \overline{Y}) \to Z = X$, where $\tau(X, \overline{Y})$ is the n-type of $(\mathfrak{T}, X, \overline{Y})|_I$.
(2) I is a D-interval for φ, X, $\overline{Y}$ if it is not a U-interval.
(3) In the special case of $I = \{u \mid u \geq z\}$ we say that the subtree $\mathfrak{T}_z$ is a U-tree or D-tree, respectively, and further say that z is a U-node or D-node for $\varphi, X, \overline{Y}$.
(4) The set of D-nodes for $\varphi, X, \overline{Y}$ is denoted $D(X)$.
(5) An infinite path P is called a D-path for $\varphi, X, \overline{Y}$ if every $v \in P$ is a D-node for $\varphi, X, \overline{Y}$, i.e. if $P \subseteq D(X)$.

Again, the name "U-interval" attests to the fact that the set X in question is *uniquely* determined by its type on a given interval, as opposed to "D-intervals" offering two (or more) distinct choices for X with the same type on the interval, thus (at least) *doubling* the total number of choices for X over the entire domain. Whenever φ and $\overline{Y}$ are clear from the context we will write e.g. "D-interval for X" instead of "D-interval for $\varphi, X, \overline{Y}$", and similarly for the other notions above.

It is worth noting that each set $D(X)$ is prefix-closed since whenever $\mathfrak{T}_v$ is a D-tree and $u < v$, then $\mathfrak{T}_v$ is a subtree of $\mathfrak{T}_u$ and hence, by composition, $\mathfrak{T}_u$ is a D-tree as well. Thus $D(X)$ induces a tree whose infinite paths are precisely the D-paths for X.

Each of the notions introduced in Definition 7.5 can be formalized in MSO. Let us start by constructing the formula $\mathrm{DINT}_\varphi(I, X, \overline{Y})$, expressing that I is a D-interval for $\varphi, X, \overline{Y}$. By Lemma 1.10, the set of n-types $H_{n,l+m+1}$ is finite and can be computed. Take the formula

$$\psi_{\mathrm{eqtp}}(X, Z, \overline{Y}) \;=\; \bigwedge_{\tau \in H_{n,l+m+1}} \tau(X, \overline{Y}) \leftrightarrow \tau(Z, \overline{Y})$$

expressing that X and Z have the same n-type (on the tree at large), and let $\psi_{\mathrm{eqtp}}^{\mathrm{rel}}(X, Z, \overline{Y}, I)$ be the relativization of $\psi_{\mathrm{eqtp}}(X, Z, \overline{Y})$ to an interval I, thus asserting that X and Z have the same n-type on I. $\mathrm{DINT}_\varphi(I, X, \overline{Y})$ can now be written as

$$\varphi(X, \overline{Y}) \;\wedge\; \exists Z(\psi_{\mathrm{eqtp}}^{\mathrm{rel}}(X, Z, \overline{Y}, I) \;\wedge\; X \cap I \neq Z \cap I).$$

Using $\mathrm{DINT}_\varphi(I, X, \overline{Y})$ one can build the formula $\mathrm{DNODE}_\varphi(v, X, \overline{Y})$ and the formula $\mathrm{DPATH}_\varphi(P, X, \overline{Y})$ expressing, respectively, that v is a D-node and that P is a D-path for $\varphi, X, \overline{Y}$. One can also construct a formula $\mathrm{DSET}_\varphi(D, X, \overline{Y})$ which holds if and only if $D = D(X)$.

The following lemma is the first step in eliminating the $\exists^{\aleph_1}$ quantifier from MSO over trees. The three cases are depicted in Figure 7.2.

Lemma 7.6. *Let $\mathfrak{T}$ be a tree and $\varphi(X, \overline{Y})$ an MSO-formula. Then, for every tuple of subsets $\overline{V}$ of $\mathfrak{T}$,*

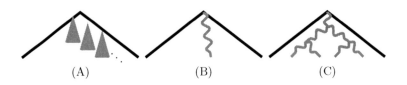

(A) (B) (C)

Fig. 7.2. The three conditions

$$\mathfrak{T} \models \exists^{\aleph_1} X \; \varphi(X, \overline{V})$$

if and only if one of the following conditions is satisfied.

A. *There is a set U satisfying $\mathfrak{T} \models \varphi(U, \overline{V})$ and there is an infinite antichain A of D-nodes for $\varphi, U, \overline{V}$.*
B. *There is an infinite branch B, which is a D-path for uncountably many U satisfying $\mathfrak{T} \models \varphi(U, \overline{V})$.*
C. *There are uncountably many branches B in $\mathfrak{T}$, each of which is a D-path for some U satisfying $\mathfrak{T} \models \varphi(U, \overline{V})$.*

Proof. Note that over finitely branching trees, where König's Lemma applies, Condition A implies Condition B and is enlisted here for deductive reasons only.

On the one hand, A is arguably the most natural and easily expressible condition sufficient for the existence of continuum many sets U satisfying $\mathfrak{T} \models \varphi(U, \overline{V})$. To see that, let U and A be as in A and let $I = \{w \in \mathfrak{T} \mid \neg\exists v \, (v \in A \wedge v < w)\}$ be the set of all nodes which are not below any of the nodes of A. Then $\mathfrak{T}$ can be decomposed with $(I, <)$ as index structure as $\mathfrak{T} = \sum_{w \in I \setminus A}[w] + \sum_{w \in A} \mathfrak{T}_w$. Here $[w]$ denotes a tree consisting of a single node bearing the same labels as w in $\mathfrak{T}$. We apply Theorem 7.4 to this decomposition. Given that $\mathfrak{T} \models \varphi(U, \overline{V})$, we can ascertain that $\mathfrak{T} \models \varphi(U', \overline{V})$ for every U' such that $U' \cap (I \setminus A) = U \cap (I \setminus A)$ and $\mathrm{Tp}^n(\mathfrak{T}_w, U', \overline{V}) = \mathrm{Tp}^n(\mathfrak{T}_w, U, \overline{V})$ for all $w \in A$. By the choice of A, such a U' can be independently chosen either to coincide or not to coincide with U on each subtree $\mathfrak{T}_w$ with $w \in A$ without changing its type. Hence there are continuum many different such U' and A is an antichain of D-nodes for every such U'. In a (finitely branching) tree with U and A fulfilling Condition A there is also, by König's Lemma, an infinite branch B such that $\mathfrak{T}_v \cap A$ is infinite for all $v \in B$. In particular, B is a D-path for each U' obtained from U as above, implying Condition B.

On the other hand, $\neg A$ amounts to saying that for each U satisfying $\varphi(U, \overline{V})$ the set $D(U)$ induces a tree comprised of only finitely many branches. In particular, that there are only finitely many infinite D-paths for each such U.

Condition B explicitly requires the existence of uncountably many sets satisfying $\varphi(X, \overline{V})$, so it too is sufficient for $\exists^{\aleph_1} X \; \varphi(X, \overline{V})$ to hold. Hence it remains to be shown that when B fails then C is both sufficient and necessary.

Assuming B does not hold in some $\mathfrak{T}$ then, as we have seen, A fails too and therefore there are only finitely many infinite D-paths for each U satisfying

$\mathfrak{T} \models \varphi(U, \overline{V})$. Also by the failure of B, every branch is a D-path for at most countably many U satisfying $\mathfrak{T} \models \varphi(U, \overline{V})$. It follows that for every such set U the collection $\{U' \mid D(U') = D(U),\ \mathfrak{T} \models \varphi(U', \overline{V})\}$ is finite or countable. Indeed, this is clear from the above whenever $D(U)$ contains an infinite D-path. If on the other hand $D(U)$ is finite then U is fully determined by $U \cap D(U)$ and the n-types of all those U-nodes that are successors of some D-node, which only allows for a finite number of choices of U given that $\mathfrak{T}$ is finitely branching.

Thus, we have established that whenever B fails in some $\mathfrak{T}$ then: there are uncountably many U satisfying $\mathfrak{T} \models \varphi(U, \overline{V})$ if and only if there are uncountably many sets $D(U)$ with $\mathfrak{T} \models \varphi(U, \overline{V})$ if and only if Condition C holds. (The last "only if" holds because in that case each relevant $D(U)$ contains only finitely many branches.)   □

We remark that Lemma 7.6 fails for infinitely branching trees. Consider a tree of depth one with the root r having countably many successor nodes and the formula $\varphi(X, Y) = X \subseteq Y$ and fix a set V of successor nodes. Then $D(X) \subseteq \{r\}$ for every X satisfying $\varphi(X, V)$, hence conditions A, B and C all fail. Note that over infinitely branching trees even the predicate $\mathrm{Inf}(X)$, meaning that the set X is infinite, cannot be expressed in pure MSO.

Let us note again that if Condition A holds then there are in fact continuum many sets X satisfying the formula $\varphi(X, \overline{Y})$. Condition A can be directly formalized in MSO(Inf), hence, over (finitely branching) trees, also in MSO as follows:

$$\psi_{\mathsf{A}}(\overline{Y}) = \exists U\, \exists A \left(\varphi(U, \overline{Y}) \wedge \mathrm{Inf}(A) \wedge \mathrm{antichain}(A) \wedge \right.$$
$$\left. \left(\forall w \in A\ \mathrm{DNODE}_\varphi(w, U, \overline{Y}) \right) \right),$$

where $\mathrm{antichain}(A) = \forall x, y \in A\ \neg(x < y \vee y < x)$.

7.2 Structure of D-Paths for Uncountably Many Sets

In this section, we show that a branch B is a witness for Condition B if and only if this branch satisfies a disjunction of three sub-conditions: Ba, Bb and Bc. Moreover, if both Condition A and Condition C fail, then already the sub-conditions Ba and Bc are sufficient. Finally, we express both Ba and Bc in MSO and show, that in fact both these sub-conditions guarantee the existence of continuum many sets X satisfying the formula $\varphi(X, \overline{Y})$ in consideration. As in the previous section, we fix an MSO formula $\varphi(X, \overline{Y})$ with $1 + m$ free variables and of quantifier rank n.

Consider the formula $\psi(X, \overline{Y}, P)$ stating that P is an infinite D-path for X and that $\varphi(X, \overline{Y})$ holds.

$$\psi(X, \overline{Y}, P) = \mathrm{DPATH}_\varphi(P, X, \overline{Y}) \wedge \mathrm{Inf}(P) \wedge \varphi(X, \overline{Y})$$

Note that a branch B witnesses Condition B in a tree $\mathfrak{T}$ if and only if $\mathfrak{T} \models \exists^{\aleph_1} U \, \psi(U, \overline{Y}, B)$. To break up Condition B for a given branch B we therefore apply the Composition Theorem for the formula ψ with the decomposition $\mathfrak{T} = \sum_{w \in B} \mathfrak{T}_{w \setminus B}$ along that branch. To that end, assuming that l labels occur in $\mathfrak{T}$ (and φ), we fix r as the number of $\mathrm{qr}(\psi)$-types in $l + m + 2$ variables, which we enumerate as $\tau_1, \ldots, \tau_r$. Then Theorem 7.4 yields a formula θ such that

$$\mathfrak{T} \models \psi(X, \overline{Y}, B) \quad \Longleftrightarrow \quad (B, <) \models \theta(P_1, \ldots, P_r) \tag{7.1}$$

with $P_i = \{w \in B \mid (\mathfrak{T}_{w \setminus B}, X, \overline{Y}, \{w\}) \models \tau_i\}$ for each $i \in \{1, \ldots, r\}$. Note that we use the expansion of $\mathfrak{T}_{w \setminus B}$ by $\{w\}$ as w is the only element of $\mathfrak{T}_{w \setminus B}$ that belongs to B.

With this reformulation it is clear that a branch B witnesses Condition B in a tree $\mathfrak{T}$ if and only if there are uncountably many different $\overline{P}$ satisfying θ, or some $\overline{P}$ satisfying θ has uncountably many X corresponding to it. Taking advantage of the fact that, by virtue of the Composition Theorem, θ merely depends on ψ but not on $\mathfrak{T}$ nor the chosen branch B, we obtain the following breakdown of Condition B.

Lemma 7.7. *Let $\mathfrak{T}$ be a tree and B an infinite branch in $\mathfrak{T}$. There are uncountably many $X \subseteq \mathfrak{T}$ satisfying the formula $\psi(X, \overline{Y}, B)$ in $\mathfrak{T}$ if and only if one of the following sub-conditions holds.*

Ba. *There exists a set X such that $\mathfrak{T}_{w \setminus B}$ is a D-interval for $\varphi, X, \overline{Y}$ for infinitely many $w \in B$.*

Bb. *There exists a set X satisfying ψ and a $w \in B$ so that*

$$\mathfrak{T}_{w \setminus B} \models \exists^{\aleph_1} X' \, \tau_i(X', \overline{Y} \cap \mathfrak{T}_{w \setminus B}, \{w\}),$$

where $\tau_i = \mathrm{Tp}^{\mathrm{qr}(\psi)}(\mathfrak{T}_{w \setminus B}, X, \overline{Y}, \{w\})$ for all $i \in \{1, \ldots, r\}$.

Bc. *It holds that*

$$(B, <) \models \exists^{\aleph_1} \overline{P} \left(\theta(\overline{P}) \wedge \bigwedge_{i=1}^{r} P_i \subseteq Q_i \wedge \forall x \bigvee_{i=1}^{r} \left(x \in P_i \wedge \bigwedge_{j \neq i} x \notin P_j \right) \right)$$

where for each $i \in \{1, \ldots, r\}$, Q_i is the set of nodes on the branch B in which the type τ_i is satisfied by some set X, i.e.

$$Q_i = \{w \in B \mid \mathfrak{T}_{w \setminus B} \models \exists X \, \tau_i(X, \overline{Y} \cap \mathfrak{T}_{w \setminus B}, \{w\})\}.$$

Proof. Recall that by (7.1) we have $\mathfrak{T} \models \psi(X, \overline{Y}, B)$ if and only if $(B, <) \models \theta(P_1, \ldots, P_r)$. We consider two cases.

Case 1: There exists a tuple $\overline{P}$ such that $(B, <) \models \theta(\overline{P})$ and there are uncountably many sets X for which $P_i = \{w \in B \mid (\mathfrak{T}_{w \setminus B}, X, \overline{Y}, \{w\}) \models \tau_i\}$ for each $i \in \{1, \ldots, r\}$.

In this case the branch B witnesses Condition B, so we only need to show that one of the sub-conditions holds. Consider a set X_0 satisfying $\psi(X_0, \overline{Y}, B)$ and having qr(ψ)-types on $\mathfrak{T}_{w \backslash B}$ for all $w \in B$ as described by $\overline{P}$. Assume that sub-condition (Ba) does not hold. Then the segment $\mathfrak{T}_{w \backslash B}$ is a U-interval for $\varphi, X_0, \overline{Y}$ for all but finitely many $w \in B$. Observe that qr(ψ) $\geq$ qr(φ). Therefore all of the uncountably many sets X that induce $\overline{P}$, i.e. have the same qr(ψ)-type as X_0 on each segment $\mathfrak{T}_{w \backslash B}$, must be equal to X_0 on all but finitely many $\mathfrak{T}_{w \backslash B}$. Therefore there is a $w \in B$ for which there are uncountably many different X having the same qr(ψ)-type as X_0 on $\mathfrak{T}_{w \backslash B}$, and thus Condition (Bb) is satisfied.

Case 2: For each tuple $\overline{P}$ such that $(B, <) \models \theta(\overline{P})$ there are only countably many sets X for which $P_i = \{w \in B \mid (\mathfrak{T}_{w \backslash B}, X, \overline{Y}, \{w\}) \models \tau_i\}$.

In this case, we show that Condition (Bc) is both necessary and sufficient for the existence of uncountably many sets X satisfying ψ.

Necessity of Condition (Bc).

As a direct consequence of (7.1) and the condition of this case, if there are uncountably many sets X satisfying ψ then there are uncountably many corresponding tuples $\overline{P}$ for which $(B, <) \models \theta(\overline{P})$. Each P_i induced by some X as in (7.1) is, by definition, the set of w's for which $(\mathfrak{T}_{w \backslash B}, X, \overline{Y}, \{w\}) \models \tau_i$. So for every $w \in P_i$ we have, in particular, that $\mathfrak{T}_{w \backslash B} \models \exists X \tau_i(X, \overline{Y} \cap \mathfrak{T}_{w \backslash B}, \{w\})$. Thus $P_i \subseteq Q_i$ for every i. Since Hintikka formulas are mutually exclusive, the P_i's are pairwise disjoint. This guarantees that the remaining conjunct $\forall x (\bigvee_{i=1}^{r} (x \in P_i \wedge \bigwedge_{s \neq r} x \notin P_s)$ of Condition (Bc) is also satisfied, and therefore Condition (Bc) holds.

Sufficiency of Condition (Bc).

By definition of the sets Q_i, for each $w \in Q_i$ there is a subset $X_{w,i} \subseteq \mathfrak{T}_{w \backslash B}$ such that $\mathfrak{T}_{w \backslash B} \models \tau_i(X_{w,i}, \overline{Y}, \{w\})$. Assuming that Condition (Bc) holds, let $\mathcal{P}$ be the uncountable set of tuples $\overline{P}$ that witness this condition. For each such tuple $\overline{P}$ and each $w \in B$ the last conjunct of Condition (Bc) guarantees that there is a unique $i = i(w, \overline{P})$ for which $w \in P_i$. Let $X_{\overline{P}} = \bigcup_{w \in B} X_{w, i(w, \overline{P})}$. Since $P_i \subseteq Q_i$, the tuple $\overline{P}$ describes indeed the types of the set $X_{\overline{P}}$ on the tree segments $\mathfrak{T}_{w \backslash B}$. According to (7.1) from $(B, <) \models \theta(\overline{P})$ we can infer that $\mathfrak{T} \models \psi(X_{\overline{P}}, \overline{Y}, B)$. Clearly, for distinct tuples $\overline{P}_1$ and $\overline{P}_2$ the sets $X_{\overline{P}_1}$ and $X_{\overline{P}_2}$ are also distinct. Therefore $\{X_{\overline{P}} \mid \overline{P} \in \mathcal{P}\}$ constitutes an uncountable family of sets satisfying ψ. $\qquad \square$

Observe that (Ba) already subsumes A in the sense that if Condition A holds then there is a branch satisfying (Ba). Also observe that Condition (Bb) is itself just another instance of our initial problem. It is important to note, however, that the above cases classify conditions under which an *individual branch* may satisfy B. At closer inspection we find that if no branch satisfies either (Bc) or (Ba) (so that in particular A fails) and moreover Condition C fails too, then (Bb) cannot hold either.

Lemma 7.8. *If over a tree $\mathfrak{T}$ both Conditions A and C fail, then Condition B implies that some branch of $\mathfrak{T}$ satisfies Condition (Ba) or Condition (Bc).*

One intuitive way to see this is that if all the conditions A, (Ba), (Bc) and C fail on a tree, and thereby also on every tree segment of that tree, then for (Bb) to hold for a proper tree segment that tree segment would have to contain a proper tree segment on which (Bb) holds, and so on indefinitely. This would ultimately trace an infinite branch witnessing (Ba), contrary to the initial assumption.

Proof. It is easy to see that if conditions A and C fail then $\mathcal{D} = \{D(X) \mid \mathfrak{T} \models \varphi(X, \overline{Y})\}$ is countable. Indeed, in the proof of Lemma 7.6 we have already remarked that the failure of A implies that each $D \in \mathcal{D}$ is a union of finitely many paths and, by definition, C holds unless there are only countably many potential D-paths in total.

If Condition B holds then there are uncountably many sets X satisfying $\varphi(X, \overline{Y})$ and thus, as $\mathcal{D}$ is countable, there is a set D such that $D = D(X)$ for uncountably many X satisfying φ. Fix such a D and consider the set of labelings $\mathcal{L} = \{\lambda^X : D \to H_{n,l+m+1} \mid D(X) = D, \mathfrak{T} \models \varphi(X, \overline{Y})\}$, where $\lambda^X(w) = \mathrm{Tp}^n(\mathfrak{T}_{w \backslash D}, X, \overline{Y})$ for all $w \in D$. We distinguish two cases.

Case 1: $\mathcal{L}$ is uncountable. Then, given that D contains only finitely many infinite paths and finitely many additional nodes, there is an infinite branch B in D such that $\{\lambda|_B \mid \lambda \in \mathcal{L}\}$ is uncountable. Observe that $\lambda^X(w) = \mathrm{Tp}^n(\mathfrak{T}_{w \backslash B}, X, \overline{Y})$ for all but finitely many nodes $w \in B$. Also observe that, since $\mathrm{qr}(\psi) \geq n$, each $\mathrm{qr}(\psi)$-type on the variables $X, \overline{Y}, B$ induces a unique n-type on the variables $X, \overline{Y}$. So there are necessarily uncountably many different partitions $\overline{P}^X = \langle P_1^X, \ldots P_r^X \rangle$ of B

$$P_j^X = \{w \in B \mid \mathrm{Tp}^{\mathrm{qr}(\psi)}(\mathfrak{T}_{w \backslash B}, X, \overline{Y}, \{w\}) = \tau_j\} \text{ for } j \in \{1, \ldots, r\},$$

with $D(X) = D$ and X satisfying φ. Using (7.1) we can check that Condition (Bc) is met for the branch B.

Case 2: $\mathcal{L}$ is countable. Then there is a type labeling $\lambda : D \to H_{n,l+m+1}$ such that $\lambda = \lambda^X$ for uncountably many X satisfying φ and having $D(X) = D$. Suppose that Condition (Ba) is not satisfied for any infinite branch B in D. Then $\lambda(w)$ uniquely determines $X \cap \mathfrak{T}_{w \backslash D}$ for all but finitely many $w \in D$ and all X satisfying φ and $D(X) = D$. Thus, there exists a $w \in D$ such that there are uncountably many X as above pairwise differing on the tree segment $\mathfrak{T}_{w \backslash D}$. However, by definition, every subtree of $\mathfrak{T}_{w \backslash D}$ is a U-tree relative to each of these X, because $D(X) = D$. Because $\mathfrak{T}$ is finitely branching, i.e. $\mathfrak{T}_{w \backslash D} \setminus \{w\}$ is a finite union of such U-trees, there can be only finitely many X as above and pairwise differing on $\mathfrak{T}_{w \backslash D}$, which is a contradiction. Therefore Condition (Ba) must hold. □

Next we will construct MSO formulas $\psi_{\mathsf{Ba}}(B, \overline{Y})$ and $\psi_{\mathsf{Bc}}(B, \overline{Y})$ formalizing sub-conditions (Ba) and (Bc), respectively. By the above, we can then use the formula $\psi_{\mathsf{B}}(\overline{Y}) = \exists B(\psi_{\mathsf{Ba}}(B, \overline{Y}) \vee \psi_{\mathsf{Bc}}(B, \overline{Y}))$ in place of Condition B in Lemma 7.6.

7.2.1   Formalization of Condition Ba

Much like Condition A, (Ba) is naturally expressible in MSO(Inf) and thus, over trees, in pure MSO as well by the formula

$$\psi_{\mathsf{Ba}}(B, \overline{Y}) = \exists X \; \exists^{\aleph_0} w \; \mathrm{DINT}(T_{w \backslash B}, X, \overline{Y}),$$

where $T_{w \backslash B}$ is just a notation for the set defined by

$$x \in T_{w \backslash B} \iff w \leq x \wedge \neg \exists b \in B \, (b > w \wedge b \leq x).$$

The fact that Condition (Ba) is sufficient for the existence of continuum many sets U satisfying $\varphi(U, \overline{V})$ can be arrived at by appealing to the Composition Theorem in the same manner as for Condition A in the proof of Lemma 7.6, because the set X can be left intact or changed to another one with the same type on any of the infinitely many trees $\mathfrak{T}_{w \backslash B}$ which are D-intervals for X.

7.2.2   Formalization of Condition Bc

In order to eliminate the explicit use of the uncountability quantifier in Condition (Bc) over $(B, <) \cong (\omega, <)$, we make use of Proposition 2.5 from [59], more directly proven in the previous chapter, which states that cardinality quantifiers can be eliminated over $(\omega, <)$.

Proposition 7.9. *For every* MSO *formula* $\varphi(\overline{X}, \overline{Y})$ *there exists an effectively constructable formula* $\psi(\overline{Y})$ *such that over* $(\omega, <)$ *the following equivalence holds:*

$$\psi(\overline{Y}) \; \equiv \; \exists^{\aleph_1} \overline{X} \, \varphi(\overline{X}, \overline{Y}) \; \equiv \; \exists^{2^{\aleph_0}} \overline{X} \, \varphi(\overline{X}, \overline{Y}).$$

Applying this result to the formula on the right hand side of Condition (Bc), with $\overline{Q}$ as parameters, we obtain a formula $\vartheta(\overline{Q})$ such that Condition (Bc) holds if and only if $(B, <) \models \vartheta(\overline{Q})$, with $\overline{Q}$ as specified there.

By Proposition 7.9, if $\vartheta(\overline{Q})$ holds, then there are even continuum many sets $\overline{P}$ satisfying Condition (Bc). This in turn ensures the existence of continuum many sets X satisfying $\varphi(X, \overline{Y})$, because for each $\overline{P}$ accounted for in $\vartheta(\overline{Q})$ a corresponding X satisfying $\psi(X, \overline{Y}, B)$ can be found and this association is necessarily injective.

To formalize Condition (Bc) in MSO over the tree $\mathfrak{T}$, we first define the sets Q_i on $\mathfrak{T}$. As the set of types is computable, we can compute each τ_i and thus effectively construct the formula $\alpha_i(w, B, \overline{Y})$ expressing that w is a node on the branch B such that $\mathfrak{T}_{w \backslash B} \models \exists X \, \tau_i(X, \overline{Y} \cap \mathfrak{T}_{w \backslash B}, \{w\})$, i.e. $w \in Q_i$. Using this formula we can express Condition (Bc) as

$$\psi_{\mathsf{Bc}}(B, \overline{Y}) = \exists \overline{Q} \left(\bigwedge_{i=1}^{r} (w \in Q_i \leftrightarrow \alpha_i(w, B, \overline{Y})) \; \wedge \; \vartheta^B(\overline{Q}) \right)$$

where ϑ^B is a relativization of ϑ to the branch B.

7.3   The Full Binary Tree and the Cantor Space

In order to formalize Condition C in MSO over trees, we first analyze the problem only on the full binary tree and identify and prove the following key topological property that distinguishes counting branches from counting arbitrary sets.

On the full binary tree $\mathfrak{T}(2) = (\{0,1\}^*, \prec, S_0, S_1)$ where $\prec$ is the prefix-order and $S_i = \{0,1\}^*i$, we show that the set of branches satisfying any given MSO formula is a Borel set in the Cantor topology and hence it has the *perfect set property*: it is uncountable iff it contains a perfect subset iff it has the cardinality of the continuum. A *perfect set* is a closed set without isolated points.

7.3.1   Overview of Topological Notions

The argument we present is based on basic results of descriptive set theory and the theory of finite automata on infinite words in connection with monadic second-order logic and the Borel hierarchy of the Cantor space. Let us recall a few basic notions from descriptive set theory. A thorough introduction to descriptive set theory can be found in [67], we only mention a few basic facts.

The Cantor space is the topological space with the product topology on $\{0,1\}^\omega$. It is a Polish space with the topology generated by basic neighborhoods $w\{0,1\}^\omega$ with the prefix $w \in \{0,1\}^*$. Alternatively, it can be defined by the metric $d(\alpha,\beta) = 2^{-\min\{n\, :\, \alpha[n] \neq \beta[n]\}}$.

The hierarchy of Borel sets is generated starting from open sets, i.e. unions of basic neighborhoods, denoted $\mathbf{\Sigma}^0_1$, and closed sets, which are complements of open sets and denoted $\mathbf{\Pi}^0_1$. Further on by transfinite induction for any countable ordinal α, $\mathbf{\Sigma}^0_\alpha$ is defined as $\{\bigcup_{i \in \omega} A_i \mid \forall i\ \exists \beta_i < \alpha\ A_i \in \mathbf{\Pi}^0_{\beta_i}\}$ and the $\mathbf{\Pi}^0_\alpha$-sets are the complements of $\mathbf{\Sigma}^0_\alpha$-sets. Each class $\mathbf{\Sigma}^0_\alpha$ and $\mathbf{\Pi}^0_\alpha$ is closed under taking inverse images by continuous functions. In fact there are complete languages in each class with respect to continuous reductions.

The projective hierarchy is built on top of the Borel hierarchy, starting with $\mathbf{\Sigma}^1_0 = \mathbf{\Pi}^1_0$ as the class of Borel sets. On the first level one has the class $\mathbf{\Sigma}^1_1$ of *analytic sets*, which are projections of Borel sets, and the class $\mathbf{\Pi}^1_1$ of *co-analytic sets*, whose complements are analytic. The hierarchy is built in this manner with sets in $\mathbf{\Sigma}^1_{\alpha+1}$ being projections of $\mathbf{\Pi}^1_\alpha$-sets, and $\mathbf{\Pi}^1_{\alpha+1}$ sets being complements of $\mathbf{\Sigma}^1_\alpha$ sets.

The connection between the topological complexity of MSO-definable tree languages and the complexity of tree-automata recognizing them is well understood [85, 70]. By Rabin's complementation theorem, all MSO-definable tree languages are in $\mathbf{\Sigma}^1_2 \cap \mathbf{\Pi}^1_2$. There are $\mathbf{\Sigma}^1_1$-complete as well as $\mathbf{\Pi}^1_1$-complete regular tree languages. For instance, the set of $\{a,b\}$-labeled binary trees, which have on every path only finitely many a's, is $\mathbf{\Pi}^1_1$-complete [3, 70]. There are regular tree languages on arbitrary finite levels of the Borel hierarchy [81].

There also exist regular tree languages not contained in $\mathbf{\Sigma}_1^1 \cup \mathbf{\Pi}_1^1$, however, languages accepted by deterministic tree automata do belong to $\mathbf{\Pi}_1^1$.

This is in stark contrast to the situation of ω-regular languages, i.e. MSO-definable sets of ω-words, which are, by McNaughton's theorem, Boolean combinations of $\mathbf{\Pi}_2^0$ sets [85].

The Cantor-Bendixson Theorem states that closed subsets of a Polish space have the *perfect set property*: they are either countable or contain a perfect subset and thus have cardinality continuum. A set P is *perfect* if it is closed and if it has no isolated points, i.e. if every open neighborhood of every point $p \in P$ contains another point of P. We shall rely on the following fundamental result on Borel sets.

Proposition 7.10 ([52, Theorem 13.6]). *Every uncountable Borel subset of a Polish space contains a perfect subset.*

In fact, Souslin has proved that all analytic sets have the perfect set property [67]. It is, however, independent of ZFC whether all co-analytic sets, or all sets on higher levels of the projective hierarchy, satisfy the continuum hypothesis [67]. A key observation that our formalization will exploit is that, even though there are non-analytic sets of trees definable in MSO, sets of definable paths are Borel.

7.3.2   Definable Sets of Branches are Borel

For a sequence $\pi \in \{0,1\}^\omega$, we denote by $\mathrm{Pref}(\pi)$ the path through the full binary tree $\mathfrak{T}(2)$ that corresponds to this sequence, which formally can be identified with the set of prefixes of π. The following theorem was recently strengthened in [15].

Theorem 7.11 (MSO definable sets of branches are Borel)
Let $U_1, \ldots, U_m$ be subsets of $\mathfrak{T}(2)$ and let $\psi(X, \overline{Y})$ be an MSO formula over $\mathfrak{T}(2)$. Then the set

$$\mathcal{X} = \{\, \pi \in \{0,1\}^\omega \mid \mathfrak{T}(2) \models \psi(\mathrm{Pref}(\pi), \overline{U}) \,\}$$

of branches of the binary tree satisfying $\psi(X, \overline{U})$ is on the third level of the Borel hierarchy, in particular, it has the perfect set property.

Proof. Given a path $\pi \in \{0,1\}^\omega$ let $B = \mathrm{Pref}(\pi)$ be the corresponding infinite branch and consider the labeled tree $\mathfrak{T}^\pi = (\mathfrak{T}(2), \mathrm{Pref}(\pi), \overline{U})$, and its decomposition as a tree sum along π: $\mathfrak{T}^\pi = \sum_{v \in B} \mathfrak{T}^\pi_{v \setminus B}$. Applying the Composition Theorem to $\mathfrak{T}^\pi$ and φ we find θ such that

$$\mathfrak{T}(2) \models \varphi(\mathrm{Pref}(\pi), \overline{U}) \iff \sum_{v \in B} \mathfrak{T}^\pi_{v \setminus B} \models \varphi \iff (B, <) \models \theta(Q_1^\pi, \ldots, Q_k^\pi)$$

where $Q_r^\pi = \{v \in B \mid \mathrm{Tp}^n(\mathfrak{T}^\pi_{v \setminus B}) = \tau_r\}$ for each $r \in \{1, \ldots, k\}$ in the enumeration of appropriate types. Note that θ does not depend on π and $(B, <) \cong (\omega, <)$.

By the well-known correspondence of MSO and finite automata there is an ω-regular language $L_\theta \subseteq (\{0,1\}^k)^\omega$ consisting of precisely those ω-words representing the characteristic sequences of predicates $\overline{Q}$ on ω for which holds $(\omega, <) \models \theta(\overline{Q})$. In particular, by McNaughton's theorem, $L_\theta \in \Sigma_3^0$ [85].

Consider now the mapping f assigning to each $\pi \in \{0,1\}^\omega$ the sequence $\rho \in (\{0,1\}^k)^\omega$ with $\rho[n] = \langle Q_r^\pi(\pi|_n) \mid r \in \{1, \ldots, k\}\rangle$. Note that if $\pi|_{n+1} = \pi'|_{n+1}$ then $Q_r^\pi(\pi|_n) \leftrightarrow Q_r^{\pi'}(\pi'|_n)$ for all $r \in \{1, \ldots, k\}$, in other words, $\rho|_n = \rho'|_n$. Therefore f is continuous with respect to the Cantor topology. By the above, $\mathcal{X} = f^{-1}(L_\theta)$ and therefore also $\mathcal{X} \in \Sigma_3^0$ as claimed. $\square$

7.4 Formalizing Existence of Uncountably Many Branches

The perfect set property established in Theorem 7.11 provides an MSO-definable characterization of Condition C of Lemma 7.6 over the full binary tree with arbitrary labeling. Via interpretations, this can be extended to all (finitely branching) trees to yield the following characterization.

Proposition 7.12 (Eliminating uncountably-many-branches quantifier). *For every MSO formula $\varphi(X, \overline{Y})$ the assertion "$\exists^{\aleph_1} B\ \mathrm{branch}(B) \wedge \varphi(B, \overline{Y})$" is equivalent over all trees to the existence of a perfect set of branches B, each satisfying $\varphi(B, \overline{Y})$. The latter ensures that there are in fact continuum many such branches.*

Proof. Perfect sets of branches are of continuum cardinality, hence the condition is clearly sufficient. Conversely, Theorem 7.11 shows that over the full binary tree with arbitrary additional unary predicates this condition is also necessary. We can transfer this result to all trees as follows.

Every tree $\mathfrak{T}$ is isomorphic to some $(T, \prec, P_1, \ldots, P_l)$ where $T \subseteq \mathbb{N}^*$ is a prefix-closed subset of finite sequences of natural numbers and $\prec$ is the prefix relation. Consider the following encoding $\mu : \mathbb{N}^* \to \{0,1\}^*$

$$(n_0, n_1, \ldots, n_s) \mapsto 0^{n_0} 1 0^{n_1} 1 \ldots 0^{n_s} 1,$$

and set $S = \mu(T)$ and $Q_i = \mu(P_i)$ for each $i = 1 \ldots l$.

Given that $v \prec w$ in $\mathfrak{T}$ if and only if $\mu(v) \prec \mu(w)$ in $\mathfrak{T}(2)$, this defines an interpretation of $\mathfrak{T}$ inside $(\mathfrak{T}(2), S, Q_1, \ldots, Q_l)$. In particular, for every MSO-formula $\vartheta(\overline{X})$ over trees with l predicates,

$$\mathfrak{T} \models \vartheta(\overline{U}) \quad \Longleftrightarrow \quad (\mathfrak{T}(2), S, Q_1, \ldots, Q_l) \models \vartheta^*(\mu(\overline{U})),$$

where ϑ^* is obtained from ϑ by interpreting each P_i with Q_i and relativizing all quantifiers to subsets or elements of S.

The embedding μ induces an injective mapping μ^* of the set of infinite branches of $\mathfrak{T}$ to infinite branches of $\mathfrak{T}(2)$. It is easy to check that μ^* is continuous.

Consider the formula $\varphi(B, \overline{Y})$ defining an uncountable set $\mathcal{D}$ of branches B of $\mathfrak{T}$ with parameters $\overline{V}$. Then $\mathcal{D}^* = \{\mu^*(B) \mid B \in \mathcal{D}\}$ is an uncountable set of branches of $\mathfrak{T}(2)$, which is defined by the formula "branch$(B) \wedge \exists$ infinite $P \subseteq B \ \varphi^*(P, \mu(\overline{V}))$" over $(\mathfrak{T}(2), S, Q_1, \ldots, Q_l)$. Hence, by Theorem 7.11, $\mathcal{D}^*$ contains a perfect set of branches. The inverse image of this set under the continuous mapping μ^* is a perfect set of branches in $\mathcal{D}$. $\qquad\square$

Towards an MSO formulation, note that the collection of nodes of a perfect set of branches induces a perfect tree, and vice versa. Let perfect(P) be a formula that expresses that P is a perfect subset, i.e. that P is prefix closed and for every $u \in P$ there are incomparable $v, w > u$ such that $v \in P$ and $w \in P$.

Corollary 7.13. *Over trees, Condition* C *is expressible in MSO as*

$$\psi_{\mathsf{C}}(\overline{Y}) = \exists P \ perfect(P) \ \wedge \ \forall B \subset P$$
$$\big(branch(B) \ \rightarrow \ \exists X \ \varphi(X, \overline{Y}) \wedge \mathrm{DPATH}_\varphi(B, X, \overline{Y})\big).$$

In particular, Condition C *entails the existence of continuum many D-paths of sets* X *satisfying* $\varphi(X, \overline{Y})$.

As we have shown above, each of the conditions of Lemma 7.6 can be formalized in MSO over trees. Thus we can again state the conclusion of this lemma: $\mathfrak{T} \models \exists^{\aleph_1} X \ \varphi(X, \overline{Y})$ holds if and only if

$$\mathfrak{T} \models \psi_{\mathsf{A}}(\overline{Y}) \ \vee \ \exists B \ (\ \psi_{\mathsf{Ba}}(B, \overline{Y}) \ \vee \ \psi_{\mathsf{Bc}}(B, \overline{Y}) \ ) \ \vee \ \psi_{\mathsf{C}}(\overline{Y}).$$

Using the above, we can reduce any formula of MSO($\exists^{\aleph_1}$) to an MSO formula equivalent over the class of trees by inductively eliminating the inner-most occurrence of a cardinality quantifier. Theorem 7.1 follows. Moreover, as we have shown in the corresponding sections, each of the conditions of Lemma 7.6 implies the existence of continuum many sets X satisfying $\varphi(X, \overline{Y})$, thus Theorem 7.2 follows as well.

8

Outlook

We considered the connection between logic and games underlying model-checking procedures on finite structures and extended it to the class of automatic structures. To this end we defined a new class of hierarchical games suitable for model-checking first-order logic over automatic structures. These games can be used not only for first-order logic, but also for formulas with the regular game quantifier. Moreover, cardinality and counting quantifiers can be reduced to first-order logic on automatic structures. Thus, hierarchical games provide a way to model-check first-order logic extended with cardinality, counting and game quantification on automatic presentations.

In our basic model of hierarchical games, two coalitions with strictly opposing objectives play a game with a particular kind of imperfect information. In general multiplayer games, strictly opposing objectives of players are uncommon. Moreover, the hierarchical constraint is a technical limitation introduced to keep the problem of establishing the winners decidable. Therefore we ask which other classes or representations of games can be used for model-checking first-order logic on automatic structures. One natural way to define such games is by departing from the standard abstraction of a token moved on a state graph and allowing the players to play with more complex objects.

For example, we imagine games where players build a new graph during the game by choosing moves labeled by some simple graph rewriting rules. One promising candidate for such rules are separated hypergraph rewriting rules introduced in [24], which were recently proven to be applicable for games as well [51]. Graphs constructed using these rules are MSO-interpretable in the binary tree, and thus have a decidable MSO theory (see [12] for an overview). Following this approach, a model-checking game for a formula $\exists x \, \forall y \, R(x, y)$, with x and y represented by finite words, would start with the Verifier building an arbitrary large graph that represents the game $\forall y \, R(u, y)$ for some word u that he chooses. Then, the Falsifier continues the construction for some word w of his choice. Finally, a regular condition is checked on the graph constructed for $R(u, w)$ to determine the winner.

Such a description seems more intuitive than the definition of hierarchical games as it involves only two players with perfect information. On the other hand, it is not clear what kind of construction rules should be allowed and how to define a natural class of such games where establishing the winner is decidable. Still, it is an interesting subject for future work to find other classes of games for model-checking first-order logic on automatic, or other finitely presented structures.

Another direction is to extend hierarchical games and to use them for model-checking on larger classes of structures. One question is whether we can obtain model-checking games for tree-automatic structures in this way. We conjecture that the answer is positive and that the necessary extension is to add two new players that transcend information levels. More precisely, the moves of the new players would be visible to all other players in the game and conversely, the new players would be able to see moves of all other players as well. The intuition is that the moves of the new players correspond to the choice of a branch in the tree when an alternating tree automaton is running. This conjecture leads to another question, namely how can such games be further extended to larger classes of structures, for example to generalized-automatic ones.

Aiming at model-checking games for larger classes of structures, we see two main directions to follow. On the one hand, games on certain infinite graphs, for example on pushdown graphs, can still be solved algorithmically. Thus, one can try to use such games for model-checking. On the other hand, one may combine the games played in the syntactic setting, like dialogue games, with model-checking games played on graphs. In this way, one views a winning strategy of the Verifier in a hierarchical game for a formula φ as a description of the choices needed to build a proof of φ by induction on the structure of words used in the presentation. The game itself is then a compact description of possible choices in the proofs for both φ and $\neg\varphi$.

For these considerations to be useful, it is necessary to find efficient algorithms for establishing the winner in the particular class of games. This is a difficult task and the procedures used to prove decidability of the games are often not efficient enough for practical applications. For example, it is unclear how to solve alternating hierarchical parity games without the complex step of determinizing alternating parity automata. Still, there are reasons to hope that it is feasible to solve even complex games and that representing problems as games helps to find efficient solutions. For example, the antichain method introduced in [21] for games with imperfect information turned out to be successful in improving model-checking algorithms based on automata [25]. We believe that further work in this direction will confirm that games can both give us better understanding of the expressive power of various logics and lead to efficient algorithms with practical applications.

References

1. Abramsky, S., Ghica, D.R., Murawski, A.S., Ong, L.C.-H.: Applying Game Semantics to Compositional Software Modeling and Verification. In: Jensen, K., Podelski, A. (eds.) TACAS 2004. LNCS, vol. 2988, pp. 421–435. Springer, Heidelberg (2004) (Cited on page VII.)
2. Alur, R., Henzinger, T.A., Kupferman, O.: Alternating-time temporal logic. In: Proceedings of the 38th Symposium on Foundations of Computer Science, FOCS 1997, pp. 100–109. IEEE, Los Alamitos (1997) (Cited on page 34.)
3. Arnold, A., Duparc, J., Murlak, F., Niwiński, D.: On the topological complexity of tree languages. In: Logic and Automata: History and Perspectives of Texts in Logic and Games, vol. 2, pp. 9–29. Amsterdam University Press, Amsterdam (2007) (Cited on page 104.)
4. Azhar, S., Peterson, G., Reif, J.H.: On multiplayer non-cooperative games of incomplete information: Part 2 - lower bounds. Technical report DUKE–TR–1991–38, Duke University (1991) (Cited on page 34.)
5. Bárány, V.: Invariants of automatic presentations and semi-synchronous transductions. In: Durand, B., Thomas, W. (eds.) STACS 2006. LNCS, vol. 3884, pp. 289–300. Springer, Heidelberg (2006) (Cited on page 10.)
6. Bárány, V.: Automatic Presentations of Infinite Structures. Dissertation, RWTH Aachen (2007) (Cited on page 10.)
7. Bárány, V., Kaiser, Ł., Rabinovich, A.: Expressing Cardinality Quantifiers in Monadic Second-Order Logic over Chains. J. Symb. Logic 76(2), 603–619 (2011) (Cited on page 79.)
8. Bárány, V., Kaiser, Ł., Rabinovich, A.: Cardinality Quantifiers in MLO over Trees. In: Grädel, E., Kahle, R. (eds.) CSL 2009. LNCS, vol. 5771, pp. 117–131. Springer, Heidelberg (2009) (Cited on pages 79 and 95.)
9. Bárány, V., Kaiser, Ł., Rabinovich, A.: Expressing Cardinality Quantifiers in Monadic Second-Order Logic over Trees. Fundamenta Informaticæ 100, 1–18 (2010) (Cited on pages 79 and 95.)
10. Bárány, V., Kaiser, Ł., Rubin, S.: Cardinality and counting quantifiers on omega-automatic structures. In: Albers, S., Weil, P. (eds.) Proceedings of the 25th International Symposium on Theoretical Aspects of Computer Science, STACS 2008 (2008) (Cited on page 67.)
11. Blumensath, A.: Automatic structures. Diploma thesis, RWTH Aachen (1999) (Cited on pages X, 12, 67, 76 and 77.)

12. Blumensath, A.: Prefix-recognisable graphs and monadic second-order logic. Technical report AIB-2001-06, RWTH Aachen (2001) (Cited on page 109.)
13. Blumensath, A., Grädel, E.: Automatic Structures. In: Proceedings of 15th IEEE Symposium on Logic in Computer Science LICS 2000, pp. 51–62 (2000) (Cited on page VIII.)
14. Blumensath, A., Grädel, E.: Finite presentations of infinite structures: Automata and interpretations. Theory of Computing Systems 37, 641–674 (2004) (Cited on pages 10, 12 and 68.)
15. Bojańczyk, M., Niwiński, D., Rabinovich, A., Radziwończyk-Syta, A., Skrzypczak, M.: On the borel complexity of mso definable sets of branches. Fundamenta Informaticae 98(4), 337–349 (2010) (Cited on page 105.)
16. Bruyère, V., Hansel, G., Michaux, C., Villemaire, R.: Logic and p-recognizable sets of integers. Bull. Belg. Math. Soc. 1, 191–238 (1994) (Cited on page 10.)
17. Büchi, J.R.: On decision method in restricted second order arithmetic. In: Proceedings of the International Congress on Logic, Methodology and Philosophy of Science, pp. 1–11 (1962) (Cited on pages VIII, 6, 8 and 13.)
18. Büchi, J.R.: Decision methods in the theory of ordinals. Bulletin of the American Mathematical Society 71, 767–770 (1965) (Cited on page 13.)
19. Büchi, J.R., Siefkes, D.: The Monadic Second Order Theory of All Countable Ordinals. Lecture Notes in Mathematics, vol. 328. Springer, Heidelberg (1973) (Cited on page 13.)
20. Bundy, A.: The automation of proof by mathematical induction. In: Handbook of Automated Reasoning, pp. 845–911. Elsevier and MIT Press (2001) (Cited on page VIII.)
21. Chatterjee, K., Doyen, L., Henzinger, T.A., Raskin, J.-F.: Algorithms for omega-regular games with imperfect information'. In: Ésik, Z. (ed.) CSL 2006. LNCS, vol. 4207, pp. 287–302. Springer, Heidelberg (2006) (Cited on pages 33 and 110.)
22. Colcombet, T., Löding, C.: Transforming structures by set interpretations. Logical Methods in Computer Science 3(2:4), 1–36 (2007) (Cited on pages X, 12 and 13.)
23. Colcombet, T., Niwinski, D.: On the positional determinacy of edge-labeled games. Theoretical Computer Science 352(1-3), 190–196 (2006) (Cited on page 64.)
24. Courcelle, B., Engelfriet, J., Rozenberg, G.: Context-free handle-rewriting hypergraph grammars. In: Ehrig, H., Kreowski, H.-J., Rozenberg, G. (eds.) Graph Grammars 1990. LNCS, vol. 532, pp. 253–268. Springer, Heidelberg (1991) (Cited on page 109.)
25. Doyen, L., Raskin, J.-F.: Improved algorithms for the automata-based approach to model-checking. In: Grumberg, O., Huth, M. (eds.) TACAS 2007. LNCS, vol. 4424, pp. 451–465. Springer, Heidelberg (2007) (Cited on page 110.)
26. Dziembowski, S., Jurdziński, M., Walukiewicz, I.: How much memory is needed to win infinite games? In: Proceedings of the 12th Annual IEEE Symposium on Logic in Computer Science, LICS 1997, pp. 99–110. IEEE Computer Society Press, Los Alamitos (1997) (Cited on page 51.)
27. Ebbinghaus, H.-D., Flum, J., Thomas, W.: Mathematical Logic. Undergraduate texts in mathematics. Springer, Heidelberg (1984) (Cited on page 1.)
28. Ehrenfeucht, A.: Decidability of the theory of the linear ordering relation. Notices of the American Mathematical Society 6, 268 (1959) (Cited on page 13.)
29. Ehrenfeucht, A.: An application of games to the completeness problem for formalized theories. Fundamenta Mathematicae 44, 241–248 (1961) (Cited on page 13.)

30. Ehrensberger, J., Zinn, C.: Dialog: A system for dialogue logic. In: McCune, W. (ed.) CADE 1997. LNCS, vol. 1249, pp. 446–460. Springer, Heidelberg (1997) (Cited on page VII.)
31. Allen Emerson, E., Jutla, C.S.: Tree automata, mu-calculus and determinacy (extended abstract). In: Proceedings of the 32nd Symposium on Foundations of Computer Science, FoCS 1991, pp. 368–377. IEEE, Los Alamitos (1991) (Cited on pages 31 and 49.)
32. Gale, D., Stewart, F.M.: Infinite games with perfect information. In: Contributions to the Theory of Games II. Annals of Mathematical Studies, vol. 28, pp. 245–266. Princeton University Press, Princeton (1953) (Cited on page 19.)
33. Gottlob, G.: Relativized logspace and generalized quantifiers over finite ordered structures. Journal of Symbolic Logic 62(2), 545–574 (1997) (Cited on page IX.)
34. Grädel, E., Kaiser, Ł.: What kind of memory is needed to win infinitary Muller games? In: Proceedings of the 7th Augustus de Morgan Workshop on Interactive Logic, Games and Social Software of Texts in Logic and Games, vol. 1, pp. 89–116. Amsterdam University Press, Amsterdam (2007) (Cited on page 47.)
35. Grädel, E., Walukiewicz, I.: Positional determinacy of games with infinitely many priorities. Logical Methods in Computer Science (2006)   (Cited on pages 51, 52 and 54.)
36. Gurevich, Y.: Monadic theory of order and topology, I. Israel Journal of Mathematics 27, 299–319 (1977) (Cited on page 86.)
37. Gurevich, Y.: Modest theory of short chains I. Journal of Symbolic Logic 44, 481–490 (1979) (Cited on pages 13 and 14.)
38. Gurevich, Y.: Monadic second-order theories. In: Barwise, J., Feferman, S. (eds.) Model-Theoretical Logics, pp. 479–506. Springer, Heidelberg (1985) (Cited on page 13.)
39. Gurevich, Y., Harrington, L.: Trees, automata and games. In: Proceedings of the 14th Annual ACM Symposium on Theory of Computing, STOC 1982, pp. 60–65. ACM, New York (1982) (Cited on pages IX, 47 and 50.)
40. Gurevich, Y., Magidor, M., Shelah, S.: The monadic theory of ω_2. Jounal of Symbolic Logic 48, 387–398 (1983) (Cited on pages 13 and 14.)
41. Gurevich, Y., Shelah, S.: Modest theory of short chains II. Journal of Symbolic Logic 44, 491–502 (1979) (Cited on pages 13 and 14.)
42. Hintikka, J.: Distributive normal forms in the calculus of predicates. Acta Philosophica Fennica 6 (1953) (Cited on page 14.)
43. Hintikka, J.: Language-games for quantifiers. Studies in Logical Theory, American Philosophical Quarterly Monograph Series 2, 46–72 (1986)   (Cited on page VII.)
44. Hjörth, G., Khoussainov, B., Montalbán, A., Nies, A.: Borel structures (2007) (manuscript) (Cited on page 77.)
45. Hodgson, B.R.: Décidabilité par automate fini. Ann. sc. math. Québec 7(1), 39–57 (1983) (Cited on pages VIII, 10 and 68.)
46. Janin, D., Walukiewicz, I.: On the expressive completeness of the propositional mu-calculus with respect to monadic second order logic. In: Sassone, V., Montanari, U. (eds.) CONCUR 1996. LNCS, vol. 1119, pp. 263–277. Springer, Heidelberg (1996) (Cited on page 32.)
47. Jurdziński, M.: Deciding the winner in parity games is in UP ∩ co-UP. Information Processing Letters 68(3), 119–124 (1998) (Cited on pages VII and 32.)
48. Jurdziński, M.: Small progress measures for solving parity games. In: Reichel, H., Tison, S. (eds.) STACS 2000. LNCS, vol. 1770, pp. 290–301. Springer, Heidelberg (2000) (Cited on pages VII and 32.)

49. Jurdzinski, M., Paterson, M., Zwick, U.: A deterministic subexponential algorithm for solving parity games. In: Proceedings of the Seventeenth Annual ACM-SIAM Symposium on Discrete Algorithms, SODA 2006, pp. 117–123. ACM Press, New York (2006) (Cited on pages VII and 32.)

50. Kaiser, Ł.: Game quantification on automatic structures and hierarchical model checking games. In: Ésik, Z. (ed.) CSL 2006. LNCS, vol. 4207, pp. 411–425. Springer, Heidelberg (2006) (Cited on page 29.)

51. Kaiser, Ł.: Synthesis for structure rewriting systems. In: Královič, R., Niwiński, D. (eds.) MFCS 2009. LNCS, vol. 5734, pp. 415–427. Springer, Heidelberg (2009) (Cited on page 109.)

52. Kechris, A.S.: Classical Descriptive Set Theory. Graduate Texts in Mathematics. Springer, Heidelberg (1995) (Cited on page 105.)

53. Khoussainov, B., Nerode, A.: Automatic presentations of structures. In: Leivant, D. (ed.) LCC 1994. LNCS, vol. 960, pp. 367–392. Springer, Heidelberg (1995) (Cited on pages VIII, 10 and 68.)

54. Khoussainov, B., Rubin, S., Stephan, F.: Definability and regularity in automatic presentations of subsystems of arithmetic. Tech. rep., University of Auckland (2003) (Cited on page 10.)

55. Khoussainov, B., Rubin, S., Stephan, F.: Definability and Regularity in Automatic Structures. In: Diekert, V., Habib, M. (eds.) STACS 2004. LNCS, vol. 2996, pp. 440–451. Springer, Heidelberg (2004) (Cited on pages 10, 69 and 75.)

56. Kolaitis, P.G.: Game quantification. In: Barwise, J., Feferman, S. (eds.) Model-Theoretical Logics, pp. 365–422. Springer, Heidelberg (1985) (Cited on pages 17 and 20.)

57. Kuske, D.: Is Ramsey's Theorem omega-automatic? In: Proceedings of the 27th International Symposium on Theoretical Aspects of Computer Science, STACS 2010, pp. 537–548 (2010) (Cited on page 77.)

58. Kuske, D., Lohrey, M.: First-order and counting theories of ω-automatic structures. In: Aceto, L., Ingólfsdóttir, A. (eds.) FOSSACS 2006. LNCS, vol. 3921, pp. 322–336. Springer, Heidelberg (2006) (Cited on pages 69, 70 and 75.)

59. Kuske, D., Lohrey, M.: First-order and counting theories of omega-automatic structures. Journal of Symbolic Logic 73, 129–150 (2008) (Cited on pages 85, 95 and 103.)

60. Lindström, P.: First order predicate logic with generalized quantifiers. Theoria 32, 186–195 (1966) (Cited on pages IX and 67.)

61. Lorenzen, P.: Logik und agon (reprinted in "Dialogische Logik"). In: Arti del XII Congresso Internationale de Filosofia, pp. 187–194 (1958) (Cited on page VII.)

62. Lorenzen, P.: Ein dialogisches konstruktivitätskriterium. In: Infinitistic Methods, Proceedings of the Symposium on Foundations of Mathematics, Polskie Wydawnictwo Naukowe, pp. 193–200 (1961) (Cited on page VII.)

63. Lorenzen, P., Lorenz, K.: Dialogische Logik. Wissenschaftliche Buchgesellschaft (1978) (Cited on page VII.)

64. Martin, D.A.: Borel determinacy. Annals of Mathematics 102, 363–371 (1975) (Cited on pages 19 and 23.)

65. McBurney, P., Parsons, S.: Dialogue games in multi-agent systems. Informal Logic (Special Issue on Applications of Argumentation in Computer Science) 22(3), 257–274 (2002) (Cited on page VII.)

66. Miyano, S., Hayashi, T.: Alternating finite automata on ω-words. Theoretical Computer Science 32, 321–330 (1984) (Cited on page 8.)

67. Moschovakis, Y.N.: Descriptive Set Theory. Studies in Logic and the Foundations of Mathematics, vol. 100. North-Holland Publishing Company, Amsterdam (1980) (Cited on pages 95, 104 and 105.)

68. Mostowski, A.W.: Games with forbidden positions. Tech. Rep. 78, Instytut Matematyki, Uniwersytet Gdański, Poland (1991) (Cited on pages 31 and 49.)

69. Niwiński, D.: On the cardinality of sets of infinite trees recognizable by finite automata. In: Tarlecki, A. (ed.) MFCS 1991. LNCS, vol. 520, pp. 367–376. Springer, Heidelberg (1991) (Cited on page 95.)

70. Niwinski, D., Walukiewicz, I.: A gap property of deterministic tree languages. Theoretical Computer Science 1(303), 215–231 (2003) (Cited on pages 95 and 104.)

71. Perrin, D., Pin, J.-E.: Semigroups and automata on infinite words. In: Fountain, J. (ed.) Semigroups, Formal Languages and Groups, pp. 49–72. NATO Advanced Study Institute, Kluwer academic publishers (1995) (Cited on pages 8 and 9.)

72. Perrin, D., Pin, J.-E.: Infinite Words (Automata, Semigroups, Logic and Games). Elsevier, Amsterdam (2004) (Cited on page 8.)

73. Pinchinat, S., Riedweg, S.: A decidable class of problems for control under partial observation. Information Processing Letters 95(4), 454–460 (2005) (Cited on page 37.)

74. Rabin, M.O.: Decidability of second-order theories and automata on infinite trees. Transactions of the American Mathematical Society 141, 1–35 (1969) (Cited on page 6.)

75. Rabinovich, A., Thomas, W.: Decidable Theories of the Ordering of Natural Numbers with Unary Predicates. In: Ésik, Z. (ed.) CSL 2006. LNCS, vol. 4207, pp. 562–574. Springer, Heidelberg (2006) (Cited on page 13.)

76. Reif, J.H.: Universal games of incomplete information. In: Proceedings of the 11th Annual ACM Symposium on Theory of Computing, STOC 1979, pp. 288–308. ACM, New York (1979) (Cited on page 33.)

77. Reif, J.H.: The complexity of two-player games of incomplete information. Journal of Computer and Systems Sciences 29(2), 274–301 (1984) (Cited on page 33.)

78. Rubin, S.: Automatic structures. Ph.D. thesis, University of Auckland, New Zealand (2004) (Cited on page 12.)

79. Rubin, S.: Automata Presenting Structures: A Survey of the Finite String Case. Bulletin of Symbolic Logic 14(2), 169–209 (2008) (Cited on page 68.)

80. Shelah, S.: The monadic theory of order. Annals of Mathematics 102, 379–419 (1975) (Cited on pages 13, 14, 89 and 91.)

81. Skurczyński, J.: The Borel hierarchy is infinite in the class of regular sets of trees. Theoretical Computer Science 112(2), 413–418 (1993) (Cited on page 104.)

82. Thomas, W.: Automata on infinite objects. In: van Leeuwen, J. (ed.) Handbook of Theoretical Computer Science, Volume B: Formal Models and Sematics, pp. 133–192. Elsevier and MIT Press (1990) (Cited on page 1.)

83. Thomas, W.: Ehrenfeucht games, the composition method, and the monadic theory of ordinal words. In: Mycielski, J., Rozenberg, G., Salomaa, A. (eds.) Structures in Logic and Computer Science. LNCS, vol. 1261, pp. 118–143. Springer, Heidelberg (1997) (Cited on page 13.)

84. Thomas, W.: Languages, automata, and logic. In: Rozenberg, G., Salomaa, A. (eds.) Handbook of Formal Languages, vol. III, pp. 389–455. Springer, New York (1997) (Cited on page 1.)

85. Thomas, W., Lescow, H.: Logical specifications of infinite computations. In: de Bakker, J.W., de Roever, W.-P., Rozenberg, G. (eds.) REX 1993. LNCS, vol. 803, pp. 583–621. Springer, Heidelberg (1994) (Cited on pages 104, 105 and 106.)

86. Vöge, J., Jurdziński, M.: A discrete strategy improvement algorithm for solving parity games. In: Emerson, E.A., Sistla, A.P. (eds.) CAV 2000. LNCS, vol. 1855, pp. 202–215. Springer, Heidelberg (2000) (Cited on pages VII and 32.)

87. Wilke, T.: An Eilenberg Theorem for ω-languages. In: Leach Albert, J., Monien, B., Rodríguez-Artalejo, M. (eds.) ICALP 1991. LNCS, vol. 510, pp. 588–599. Springer, Heidelberg (1991) (Cited on page 8.)

88. Zielonka, W.: Infinite games on finitely coloured graphs with applications to automata on infinite trees. TCS 200(1–2), 135–183 (1998) (Cited on pages IX, 47, 62 and 66.)

Index